EXTENDING THE HORIZONS: ADVANCES IN COMPUTING, OPTIMIZATION, AND DECISION TECHNOLOGIES

OPERATIONS RESEARCH/COMPUTER SCIENCE INTERFACES SERIES

Professor Ramesh Sharda
Oklahoma State University

Prof. Dr. Stefan Voß
Universität Hamburg

Greenberg / *A Computer-Assisted Analysis System for Mathematical Programming Models and Solutions: A User's Guide for ANALYZE*
Greenberg / *Modeling by Object-Driven Linear Elemental Relations: A Users Guide for MODLER*
Brown & Scherer / *Intelligent Scheduling Systems*
Nash & Sofer / *The Impact of Emerging Technologies on Computer Science & Operations Research*
Barth / *Logic-Based 0-1 Constraint Programming*
Jones / *Visualization and Optimization*
Barr, Helgason & Kennington / *Interfaces in Computer Science & Operations Research: Advances in Metaheuristics, Optimization, & Stochastic Modeling Technologies*
Ellacott, Mason & Anderson / *Mathematics of Neural Networks: Models, Algorithms & Applications*
Woodruff / *Advances in Computational & Stochastic Optimization, Logic Programming, and Heuristic Search*
Klein / *Scheduling of Resource-Constrained Projects*
Bierwirth / *Adaptive Search and the Management of Logistics Systems*
Laguna & González-Velarde / *Computing Tools for Modeling, Optimization and Simulation*
Stilman / *Linguistic Geometry: From Search to Construction*
Sakawa / *Genetic Algorithms and Fuzzy Multiobjective Optimization*
Ribeiro & Hansen / *Essays and Surveys in Metaheuristics*
Holsapple, Jacob & Rao / *Business Modelling: Multidisciplinary Approaches — Economics, Operational and Information Systems Perspectives*
Sleezer, Wentling & Cude/*Human Resource Development And Information Technology: Making Global Connections*
Voß & Woodruff / *Optimization Software Class Libraries*
Upadhyaya et al / *Mobile Computing: Implementing Pervasive Information and Communications Technologies*
Reeves & Rowe / *Genetic Algorithms—Principles and Perspectives: A Guide to GA Theory*
Bhargava & Ye / *Computational Modeling And Problem Solving In The Networked World: Interfaces in Computer Science & Operations Research*
Woodruff / *Network Interdiction And Stochastic Integer Programming*
Anandalingam & Raghavan / *Telecommunications Network Design And Management*
Laguna & Martí / *Scatter Search: Methodology And Implementations In C*
Gosavi/ *Simulation-Based Optimization: Parametric Optimization Techniques and Reinforcement Learning*
Koutsoukis & Mitra / *Decision Modelling And Information Systems: The Information Value Chain*
Milano / *Constraint And Integer Programming: Toward a Unified Methodology*
Wilson & Nuzzolo / *Schedule-Based Dynamic Transit Modeling: Theory and Applications*
Golden, Raghavan & Wasil / *The Next Wave in Computing, Optimization, And Decision Technologies*
Rego & Alidaee/ *Metaheuristics Optimization via Memory and Evolution: Tabu Search and Scatter Search*
Kitamura & Kuwahara / *Simulation Approaches in Transportation Analysis: Recent Advances and Challenges*
Ibaraki, Nonobe & Yagiura / *Metaheuristics: Progress as Real Problem Solvers*
Golumbic & Hartman / *Graph Theory, Combinatorics, and Algorithms: Interdisciplinary Applications*
Raghavan & Anandalingam / *Telecommunications Planning: Innovations in Pricing, Network Design and Management*
Mattfeld / *The Management of Transshipment Terminals: Decision Support for Terminal Operations in Finished Vehicle Supply Chains*
Alba & Martí/ *Metaheuristic Procedures for Training Neural Networks*
Alt, Fu & Golden/ *Perspectives in Operations Research: Papers in honor of Saul Gass' 80th Birthday*

EXTENDING THE HORIZONS: ADVANCES IN COMPUTING, OPTIMIZATION, AND DECISION TECHNOLOGIES

Edited by

EDWARD K. BAKER
University of Miami

ANITO JOSEPH
University of Miami

ANUJ MEHROTRA
University of Miami

MICHAEL A. TRICK
Carnegie Mellon University

Springer

Edward K. Baker
Anito Joseph
Anuj Mehrotra
University of Miami
Florida, USA

Michael A. Trick
Carnegie Mellon University
Pennsylvania, USA

Library of Congress Control Number: 2006937293

ISBN-10: 0-387-48790-5 (HB) ISBN-10: 0-387-48793-X (e-book)
ISBN-13: 978-0-387-48790-8 (HB) ISBN-13: 978-0-387-48793-9 (e-book)

Printed on acid-free paper.

9 8 7 6 5 4 3 2 1

springer.com

Contents

Preface

This book is the special volume published in conjunction with the Tenth INFORMS Computing Society conference to be held January 3-5, 2007 in Coral Gables, Florida. The title of the book, *Extending the Horizons: Advances in Computing, Optimization, and Decision Technologies*, echoes the general theme of the conference. The volume contains 15 high quality papers that span the range of computing, optimization, and decision technologies. This volume is a carefully edited and refereed collection of papers selected from among those submitted for the conference.

The contents of this volume reflect the current research interests of many of the INFORMS Computing Society's members. The first paper in the volume, and the topic of the plenary talk at the conference, extends computing and optimization into the area of improving the global modeling of hurricanes: a decisions technology particularly apropos to the conference location in South Florida. The remainder of the volume follows in five sections. Networks and Graphs are the topics for Section II. In Section III, four different papers consider both computational and theoretical issues in Optimization. Vehicle Routing and the Traveling Salesman Problem are considered in Section IV. Section V contains papers that focus on Simulation issues, while Section VI considers innovative Decision Technologies.

We would like to thank all of those who served as referees for the papers submitted for publication in this volume and we appreciate their willingness and cooperation in making the high level of scholarship found in this volume possible. Additionally, we would like to acknowledge the School of Business at the University of Miami and the Tepper School of Business at Carnegie Mellon University for their very generous support of their time and resources.

We look forward to a very successful Tenth INFORMS Computing Society conference.

Edward Baker, Anito Joseph, Anuj Mehrotra, and Mike Trick

IMPROVING HURRICANE PREDICTION THROUGH INNOVATIVE GLOBAL MODELING

Robert Atlas,[1] Shian-Jiann Lin,[2] Bo-Wen Shen,[3] Oreste Reale,[4] and Kao-San Yeh[4]
[1] NOAA/Atlantic Oceanographic and Meteorological Laboratory, Miami, Florida; [2]NOAA/Geophysical Fluid Dynamics Laboratory, Princeton, New Jersey; [3]University of Maryland, College Park, Maryland; [4]University of Maryland, Baltimore County, Maryland

Abstract: Current global and regional models incorporating both in situ and remotely sensed observations have achieved a high degree of skill in forecasting the movement of hurricanes. Nevertheless, significant improvements in the prediction of hurricane landfall and intensification are still needed. To meet these needs, research on new observing systems, data assimilation techniques, and better models is being performed. These include the Hurricane Weather Research and Forecasting regional model development by NOAA, as well as the development of an advanced "seamless" global weather and climate model, as a collaborative project involving both NOAA and NASA. This latter model, when completed, will be used to improve short and extended range forecasts of hurricanes, as well as to determine the relationship between global climate change and long-term variations in hurricane frequency and intensity, more accurately than is possible today. As a starting point for the seamless global weather and climate model, the horizontal resolution of the previously developed finite volume General Circulation Model has been increased to 1/12° (approximately 9 km) in a series of successive steps. This was made possible by advances in both computing and optimization technologies.

Key words: Hurricane prediction; global modeling.

1. INTRODUCTION

Each year hurricanes, typhoons, and other tropical cyclones cause thousands of fatalities and tens of billions of dollars of economic losses throughout the world. Severe examples include the tropical cyclone that killed more than 300,000 people in Bangladesh in 1970, and in the United States: the Galveston Hurricane of 1900, which destroyed the city and killed between 6000 and 8000 people; Hurricane Andrew, which caused monetary losses of 26.5 billion dollars (normalized to 38 billion dollars by inflation, wealth and population changes) in 1992 (Pielke and Landsea 1998); and, most recently, Hurricane Katrina, which killed more than 1300 people and resulted in losses in excess of 100 billion dollars. Even storms of much lesser intensity can produce significant loss of life and property, presenting a daunting challenge for hurricane forecasters and the communities they serve.

Although individual years may vary, the number of hurricanes and the number of major hurricanes (defined as Category 3 or higher on the Saffir-Simpson scale) has been increasing in recent years. The year 2004 was a very active season for the North Atlantic with 15 named storms, nine of which became hurricanes, and six of which became major hurricanes. These included Hurricanes Charlie, Frances, Ivan and Jeanne, which all caused extensive damage and loss of life. The year 2005 continued this upward trend, with 28 named storms, 15 hurricanes, three Category 5 hurricanes, and four major hurricanes hitting the United States.

The reduction of losses related to hurricanes involves many complex aspects ranging from purely theoretical, observational, computational and numerical, to operational and decisional. A correct warning can lead to proper evacuation and damage mitigation, and produce immense benefits. However, over-warning can lead to substantial unnecessary costs, a reduction of confidence in warnings, and a lack of appropriate response. In this chain of information, the role played by scientific research is crucial. Within the United States, the U.S. Weather Research Program (USWRP) has been addressing this problem by coordinating research amongst federal agencies and academic institutions in order to reduce the landfall, track and intensity forecast errors, increase the warning lead time, and extend the period for which hurricane and precipitation forecasts are useful. The National Oceanic and Atmospheric Administration (NOAA), in combination with the National Aeronautics and Space Administration (NASA) and other agencies, is contributing to these efforts through observational and theoretical research to better understand the processes associated with the formation, intensification, and movement of hurricanes. This includes model and data assimilation development, Observing System Experiments (OSE) and Observing System Simulation Experiments (OSSE) designed to ascertain the value of existing observing systems and the potential of new observing systems to improve hurricane prediction, and theoretical research to improve understanding of hurricane development and evolution. In this paper, we report on some of the innovative research that is being performed to develop an advanced next-generation global model for improved hurricane prediction.

2. BACKGROUND

Numerical weather prediction (NWP) is an initial-value problem that depends upon the quality of the initial condition and the accuracy of the computer model that predicts the evolution of the weather systems (Lin et al. 2004). Since the initial conditions for the atmosphere and the ocean cannot be perfectly prescribed, the predictability of the weather is limited by the error in the initial conditions, as well as by the chaotic nature of the dynamics and physics that amplifies the initial errors (Kalnay 2003). Errors in the initial state grow rapidly in time within a given model (Atlas et al. 2005a). These are addressed by developing and deploying improved observing systems, and by the development of improved methods for assimilating these observations into the model's initial state (Atlas et al. 2005b). Beyond the techniques to improve the quality of the initial condition, it is crucial to design an atmospheric model so that it minimizes the amplification of initial errors, and this relies on our understanding of the weather phenomena, as well as computing and optimization technologies.

A major area where improvements to models can be made is in the numerical approximations to the dynamical and physical processes. The analytic equations governing the fluid-dynamical processes of the atmosphere and the ocean have been known for more than a century. It is the numerical solutions to these well-known equations that can be improved by advanced numerical algorithms and by increasing the resolution. The errors in the parameterized physical processes, however, can not be reduced by simply increasing the resolution, because some of the physical processes, such as moist convection for the formation of clouds and the associated cloud-radiation processes, are not yet sufficiently well understood and not fully described by existing equations. In particular, cumulus scales are still not predictable beyond a few hours. Increasing the resolution, however, can reduce the reliance on physical parameterizations and lead to the direct use of explicit formulation for important physical processes (e.g., cloud microphysics instead of cumulus parameterization), although simulation of precipitation still requires parameterization of the microphysical processes that govern the evolution of the droplet size spectrum. Therefore, our approach in the current and the future modeling system is to increase the resolution to the maximum extent allowed by available computer platforms, and to develop a direct, physically based approach to modeling the physical processes at that resolution.

Hurricanes are a particularly difficult task for NWP, because of their small scale, and rapidly and dramatically evolving life cycles. Hurricane life cycles often extend 5–7 days, and range over thousands of miles over the Atlantic Ocean and North America. In a typical scenario, hurricanes may start as tropical waves off the west coast of Africa, then develop and intensify as cyclones over the tropical Atlantic Ocean, reach a mature stage as they move toward the continental U.S. or Mexico, and finally weaken as they move over the land. The prediction of weather systems over such enormous temporal and spatial scales requires a global model to provide accurate lateral boundary conditions, and the violent nature and small-scale structure of hurricanes requires very high resolution for successful simulations. Subject to limited computing power in the past, hurricane prediction was mostly accomplished with regional models. Fortunately, global high-resolution modeling of the atmosphere has become practical with recent significant advances in computing technology that allow a trillion floating-point operations per second (TFLOPS) computing capacity, e.g., the Japanese Earth Simulator and the NASA Columbia supercomputer. This motivates the redesign of global models with more precise dynamical formulations and more detailed physical parameterizations, in order to simulate fine-scale processes in the global domain.

3. INNOVATIVE GLOBAL MODELING

At the present time, global and regional models incorporating both in situ and remotely sensed observations have achieved a high degree of skill in forecasting the movement of hurricanes. Nevertheless, significant improvements in the prediction of hurricane landfall and intensification are still needed. To meet these needs, research on new observing systems, data assimilation techniques, and better models is being performed. These include the Hurricane Weather Research and Forecasting regional model development by NOAA, as well as the development of an advanced "seamless" global weather and climate model, as a collaborative project involving both NOAA and NASA. This latter model, when completed, will be used to improve short and extended range forecasts of hurricanes, as well as to determine the relationship between global climate change and long-term variations in hurricane frequency and intensity, more accurately than is possible today.

The objective of the latter activity is to develop a comprehensive global model that will explicitly resolve weather and climate relevant processes, in order to improve dramatically the use of in situ and space-based observations, and the application of these observations to the understanding and prediction of weather and climate. This will require (a) non-hydrostatic atmospheric dynamics on quasi-uniform grids with resolution determined by observed process scales, (b) explicit microphysics of clouds that represent observed cloud processes and their interactions with radiation, (c) an ultra-high-resolution land surface model with process-scale dynamics, (d) an eddy-resolving ocean model, (e) coupled model evaluation and refinement based on high-resolution satellite observations, and (f) a common software environment to enable component model coupling, inclusion of observational constraints, and research community interaction.

3.1 Initial Development

The starting point for the development of an advanced next-generation global model, applicable to improved short and extended range hurricane prediction, is the finite volume General Circulation Model (fvGCM). This model was previously developed at the NASA Data Assimilation Office, and is now being used and further developed as a collaboration between NOAA and NASA. The fvGCM was designed with innovative algorithms for global high-resolution modeling of the atmosphere. The finite-volume transport scheme conserves mass locally and monotonically to ensure proper correlations among the constituents (Lin and Rood 1996). The vorticity-preserving horizontal dynamics enhances the simulation of atmospheric oscillations and vortices, which are characteristic of climate and weather phenomena (Lin and Rood 1997). The Lagrangian vertical dynamics accurately consolidates the horizontal dynamics into physically consistent three-dimensional dynamics (Lin 2004). Physical processes, such as cumulus parameterization and gravity-wave drag, are largely enhanced with emphases for high-resolution simulations; they are also modified for consistent application with the innovative finite-volume dynamics. Above all the scientific renovations, the local nature of the algorithms is the key to efficient optimization with modern distributed-memory computers (Yeh et al. 2002), which makes it possible to advance hurricane prediction to a new frontier.

The early applications of the fvGCM were found to be extremely promising, and over the past two years, we have demonstrated substantial advances in the representation of hurricanes and other weather phenomena, as the resolution of the fvGCM has been increased. Originally run at a resolution of 2° latitude by 2.5° longitude, the fvGCM (also referred to as the NASA/NCAR model) displayed an outstanding ability to simulate large scale climatic features. When the resolution was increased to 1° latitude by 1.25° longitude, the simulation of atmospheric fronts became possible, and very accurate predictions of major mid-latitude snowstorms were obtained. Upon increasing the resolution to .5° latitude by .625° longitude, the simulation of hurricanes with reasonable structure and evolution became possible. Figure 1 shows an example of hurricane evolution within a long climate simulation of the fvGCM at this resolution. The simulated hurricane in this case displays characteristic features including an eye, an eyewall (where maximum winds occur), and spiral bands of intense precipitation as it evolves in a realistic manner.

As an illustration of an actual forecast at 1/2° resolution, Figure 2 shows the track of the 1999 Hurricane Floyd predicted by the fvGCM is almost identical to the observation, except for an error in timing.

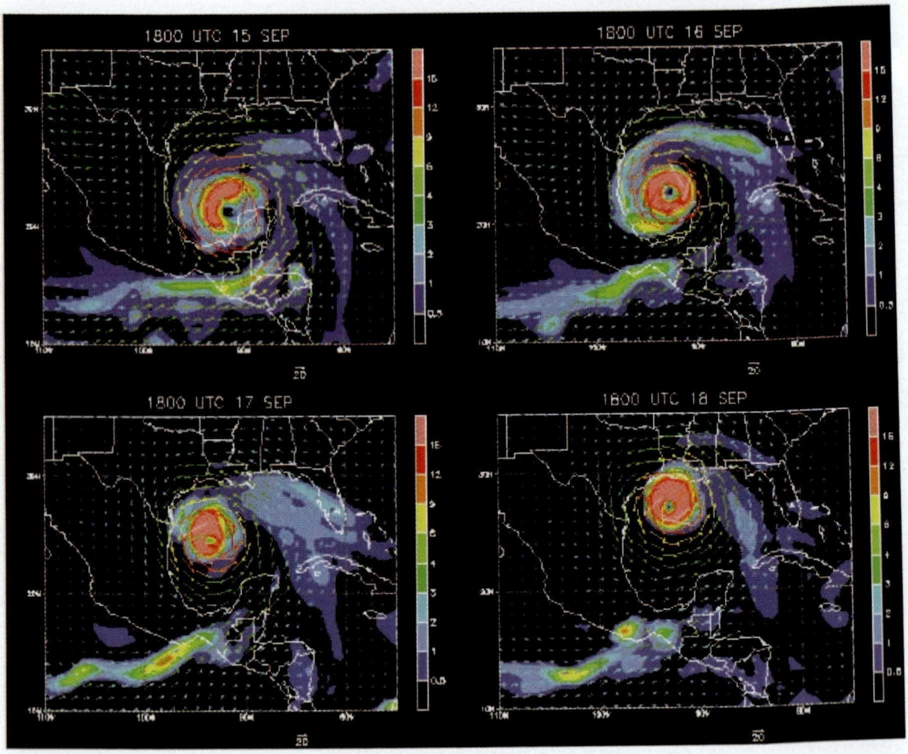

Fig. 1. A hurricane simulated by the fvGCM showing the hurricane eye, eyewall, and realistic spiral bands. The precipitation rate (mm/hr) is depicted with the color scheme on the right, and the wind (m/s) is shown with magnitude proportional to the arrow (20 m/s) at the bottom.

3.2 Impacts of Computing and Optimization Technologies

Increased horizontal resolution is crucial to the prediction of hurricanes, because their formation with organized convection has a fine-scale nature in the horizontal. The increase of horizontal resolution is, however, very expensive, and it is more so for global models which have much larger domains. Each doubling of horizontal resolution, entailed with corresponding increase of temporal resolution, typically costs 6-10 times the computing resource, depending on the type of computer, the optimization techniques, and the number of CPUs used in the application. Global modeling with 1/4° and higher resolutions is made practically possible only recently with the revolutionary advances in computing and optimization technologies. This has extended the prediction of hurricanes to a new horizon.

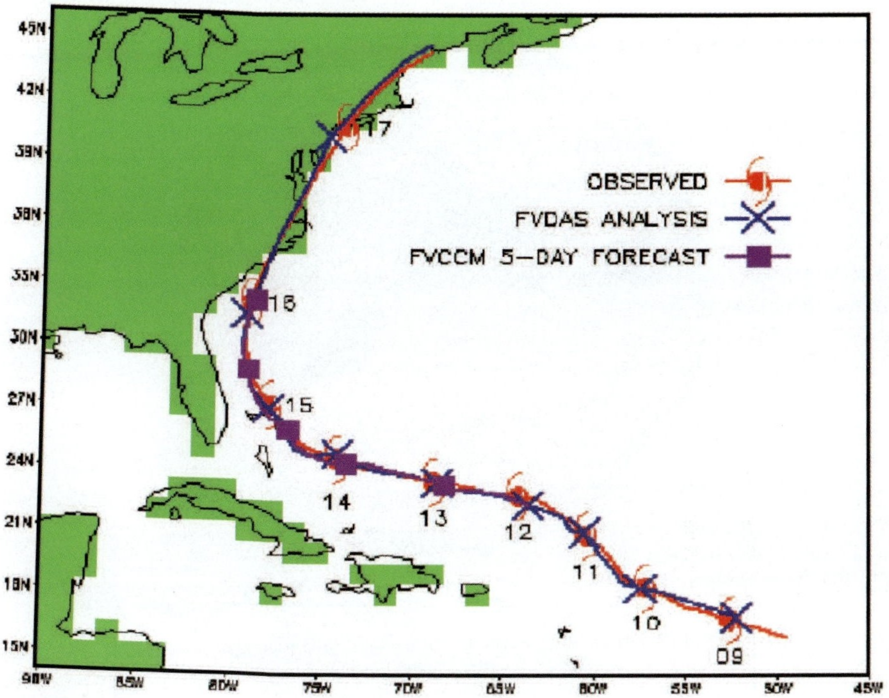

Fig. 2. Validation of the finite-volume General Circulation Model (purple squares) and Data Assimilation System (blue crosses) with the observed track of Hurricane Floyd (red spiral spots) observed by the National Hurricane Center in September 1999.

Using the NASA Columbia supercomputer, the (horizontal) resolution of the fvGCM has been successfully advanced from 1/2° to 1/4° (Atlas et al. 2005a) and then experimentally to 1/8° and 1/12°. Figure 3 illustrates how the increasing resolution can improve weather prediction in the tropics. The top panel shows that at 1/4° resolution, mesoscale structures, such as moisture filaments turning into spiral bands in the formation of hurricanes, are observed in a very realistic manner. Resolving such fine-scale features and processes can lead to more accurate prediction of both hurricane tracks and intensity. During 2004, the fvGCM was run every day in real time at 1/4° resolution in order to provide experimental forecasts for each of the hurricanes. Using 240 processors on Columbia, a five-day global forecast at 1/4° resolution was performed within 40 minutes. The bottom panel of Figure 3 shows the fvGCM five-day forecast of the evolution and landfall of Hurricane Ivan. In this forecast, the landfall of Ivan was forecast to within 56 kilometers of its observed location. For Hurricane Jeanne (not shown), the five-day forecast of landfall forecast was only 2 kilometers in error, but not all forecasts were able to achieve a similar level of accuracy. With the 1/8° resolution version of the fvGCM (Figure 4), mesoscale

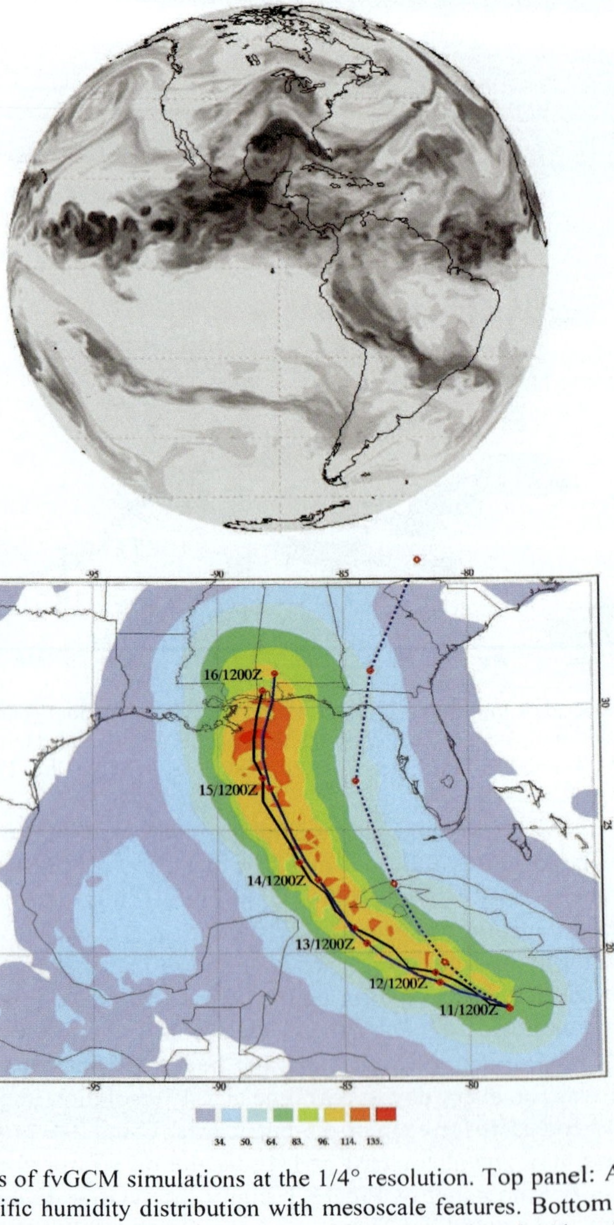

Fig. 3. Examples of fvGCM simulations at the 1/4° resolution. Top panel: A very realistic 500 mb specific humidity distribution with mesoscale features. Bottom panel: Five-day fvGCM forecast (black curve) of the 2004 Hurricane Ivan track. Also shown are the observed track (solid blue curve) and the operational forecast track based on other models (dashed blue curve). Shading represents the fvGCM-forecasted maximum sustained surface wind speed (knots), and shows the significant intensification of Hurricane Ivan from Category 1 to Category 4 prior to landfall.

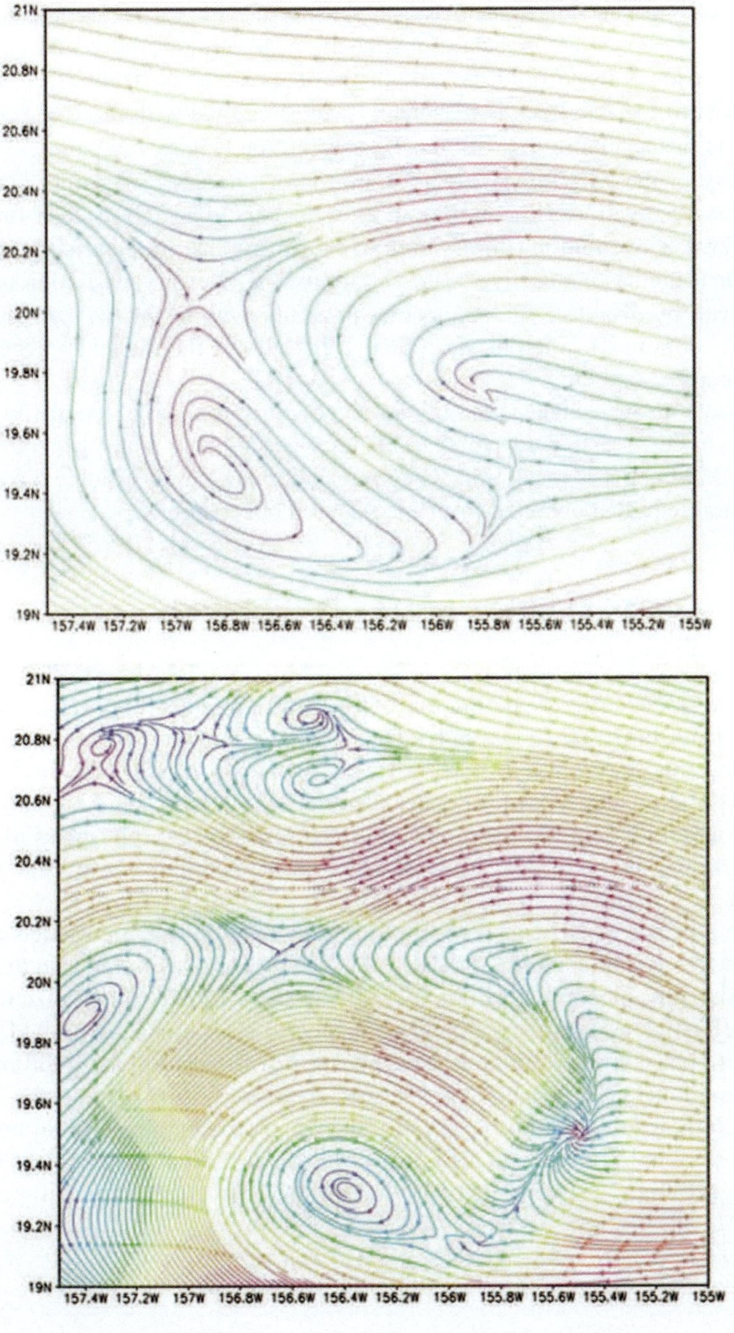

Fig. 4. Simulation of surface winds of Hawaiian wakes at 1/4° (top) and 1/8° (bottom) resolutions. (*Courtesy American Geophysical Union,* Shen et al. 2006b).

features were found to be significantly better resolved than at 1/4° resolution, and this is critical for obtaining further improvements to the prediction of hurricane intensity.

Figure 5 shows the impacts of increasing computing power on the prediction of Hurricane Katrina (2005), through the increase of horizontal resolution. It can be seen that the track is reasonably well predicted with the 1/4° resolution, and it is further improved with the 1/8° resolution (Fig. 5, top panel). The wind intensity of the 1/4° forecast (Fig. 5, bottom b), however, does not agree well with the observations (Fig. 5, bottom a); it appears to have been over forecasted with the 1/8° resolution (Fig. 5, bottom c). Noting that the internal structure of the hurricane has convective-scale variations, the intensity prediction is largely improved by disabling the convection parameterization for the 1/8° resolution (Fig. 5, bottom d) (Shen et al. 2006a). This allows the model to generate the precipitation explicitly.

Results at even higher resolution are equally promising. As an illustration, Figure 6 shows a five-day forecast of total precipitable water generated using the fvGCM at 1/12° resolution. The representation of meteorological features is comparable to that obtained from space-based observations.

4. CONCLUSIONS AND FUTURE WORK

This paper shows that advances in computing and optimization can extend the horizon of hurricane prediction through innovative global high-resolution modeling. Both hurricane track and intensity predictions can be improved by increasing model resolution in accord with the superior power of modern computers. The prediction of hurricanes also depends heavily on the quality of the initial condition input to the model, and this relies on the robustness of observing systems and the analysis techniques that assimilate these data into the model.

The advance of horizontal resolution on global atmospheric models is, however, severely subject to the model structure associated with the grid that represents the atmosphere in the computer, due to the difficulty in optimization. The parallel efficiency of many global atmospheric models is still limited by one-dimensional domain decomposition, because of the use of the traditional latitude-longitude grids, which have the meridians converged at the poles. To fundamentally resolve the optimization issue for global high-resolution modeling, it is desirable to use quasi-uniform grids, such as the cubed sphere or the geodesic grid (Fig. 7), for efficient two-dimensional domain decomposition.

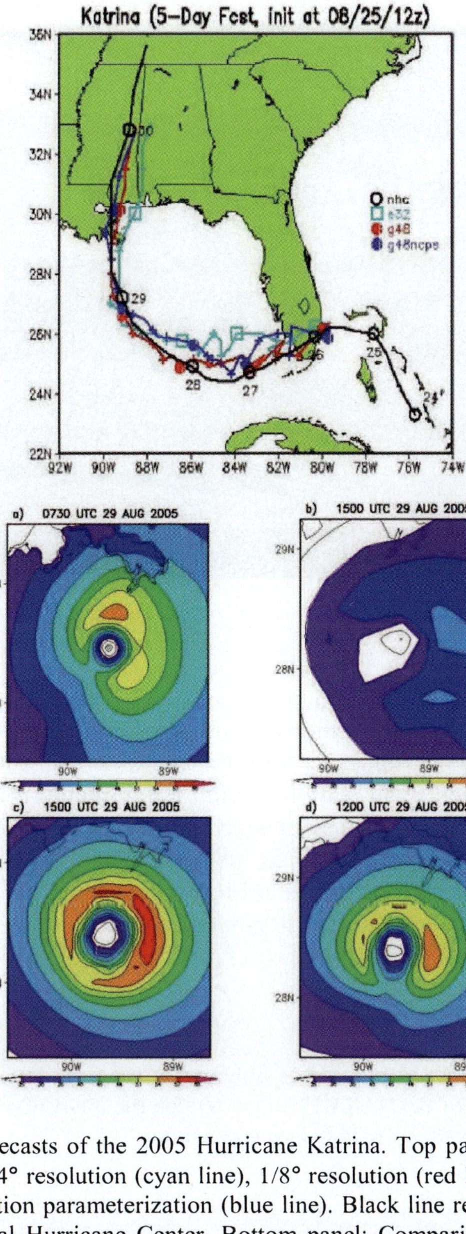

Fig. 5. Five-day forecasts of the 2005 Hurricane Katrina. Top panel: Tracks predicted by the fvGCM at 1/4° resolution (cyan line), 1/8° resolution (red line), and 1/8° resolution without convection parameterization (blue line). Black line represents the observations by the National Hurricane Center. Bottom panel: Comparison of wind intensity near the hurricane eye in a 2° × 2° box among (a) High-resolution (0.0542°) analysis at 0730 UTC AUG 29 by the Atlantic Oceanographic and Meteorological Laboratory, (b) 1/4° model forecast at 1500 UTC AUG 29, (c) 1/8° model forecast at 1500 UTC AUG 29, and (d) 1/8°-no-convection-parameterization forecast at 1200 UTC AUG 29. (*Courtesy American Geophysical Union,* Shen et al. 2006a).

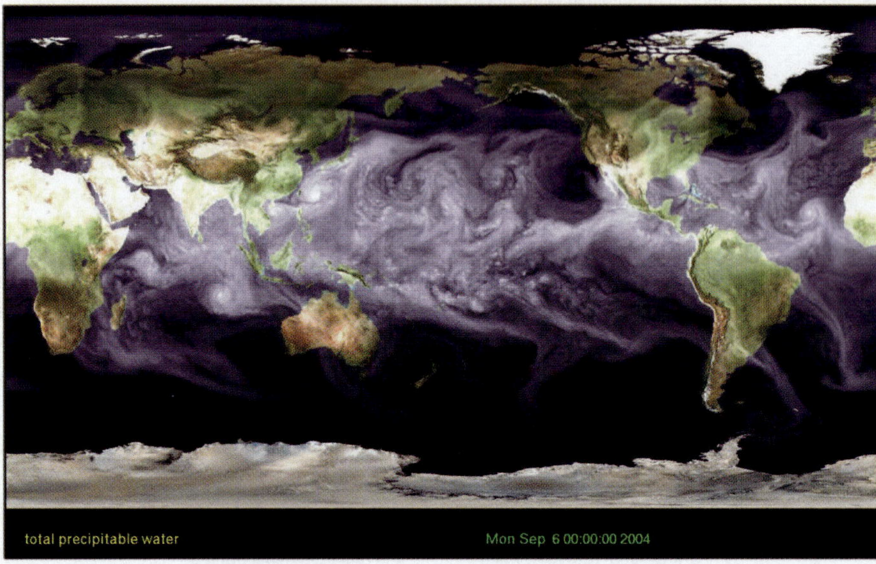

Fig. 6. Five-day forecast of total precipitable water using the 1/12° resolution version of the fvGCM.

In our future work, a non-hydrostatic version of the finite volume dynamics will be applied on quasi-uniform grids and at much higher resolution. In addition, explicit cloud microphysics will be incorporated, and the atmospheric model will be coupled to an eddy resolving ocean model. These developments will enable far more effective use of high resolution observational data and are expected to lead to further significant improvements to hurricane track and intensity prediction, as well as to the ability to determine the effect of potential climatic changes on hurricane frequency and intensity more accurately than is possible today.

Acknowledgements

This work would not have been possible without the initial support of Drs. Ghassem Asrar and Tsengdar Lee of the NASA Headquarters on the innovative model development and on the use of the NASA Columbia supercomputer.

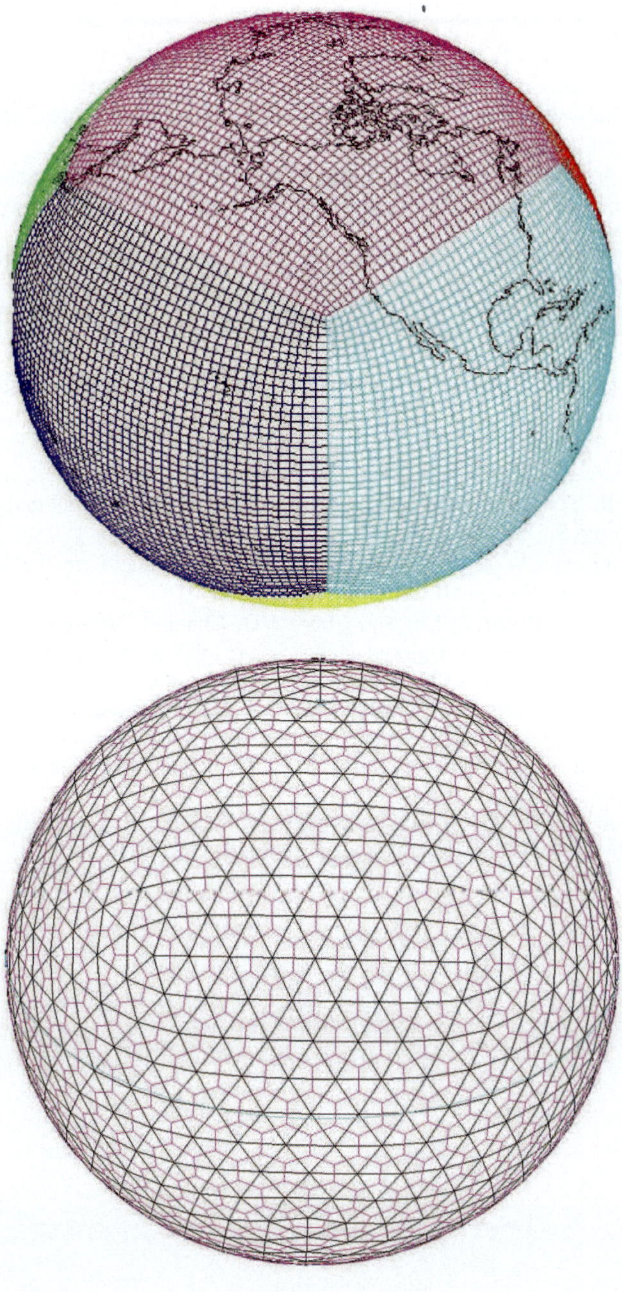

Fig. 7. Quasi-uniform grids for efficient global high-resolution modeling. Top panel: Cubed sphere. Bottom panel: Icosahedral geodesic grid.

5. REFERENCES

Atlas R, Reale O, Shen B-W, Lin S-J, Chern J-D, Putman W, Lee T, Yeh K-S, Bosilovich M, Radakovich J (2005a) Hurricane forecasting with the high-resolution NASA finite-volume General Circulation Model. Geophys Res Let 32: doi:10.1029/2004GL021513

Atlas R, Hou AY, Reale O (2005b) Application of SeaWinds scatterometer and TMI-SSM/I rain rates to hurricane analysis and forecasting. ISPRS J Photogram Remote Sens 59:233–243

Kalnay E (2003) Atmospheric modeling, data assimilation and predictability. Cambridge Univ. Press

Lin S-J (2004) A vertically Lagrangian finite-volume dynamical core for global models. Mon Wea Rev 132:2293–2307

Lin S-J, Rood RB (1996) Multidimensional flux form semi-Lagrangian transport schemes. Mon Wea Rev 124:2046–2070

Lin S-J, Rood RB (1997) An explicit flux-form semi-Lagrangian shallow-water model on the sphere. Q J Roy Met Soc 123:2477–2498

Lin S-J, Atlas R, Yeh K-S (2004) Global weather prediction and high-end computing at NASA. Comp Sci Eng 6:29–35

Pielke RA, Landsea CW (1998) Normalized hurricane damages in the United States: 1925–1995. Wea Forecast 13:621–631

Shen B-W, Atlas R, Reale O, Lin S-J, Chern J-D, Chang J, Henze C, Li J-L (2006a) Hurricane forecasts with a global mesoscale-resolving model: Preliminary results with Hurricane Katrina (2005). Geophys Res Let 33:doi:10.1029/2006GL026143

Shen B-W, Atlas R, Chern J-D, Reale O, Lin S-J, Lee T, Chang J (2006b) The 0.125 degree finite-volume general circulation model on the NASA Columbia supercomputer: Preliminary simulations of mesoscale vortices. Geophys Res Let 33: doi:10.1029/2005GL024594

Yeh K-S, Lin S-J, Rood RB (2002) Applying local discretization methods in the NASA finite-volume general circulation model. Comp Sci Eng 4:49–54

A BRANCH-AND-PRICE APPROACH FOR GRAPH MULTI-COLORING

Anuj Mehrotra
Department of Management Science
School of Business Administration
University of Miami
Coral Gables, FL 33124-8237
anuj@miami.edu

Michael A. Trick
Tepper School of Business
Carnegie Mellon University
Pittsburgh, PA 15213-3890
trick@cmu.edu

Abstract We present a branch-and-price framework for solving the graph multi-coloring problem. We propose column generation to implicitly optimize the linear programming relaxation of an independent set formulation (where there is one variable for each independent set in the graph) for graph multi-coloring. This approach, while requiring the solution of a difficult subproblem, is a promising method to obtain good solutions for small to moderate size problems quickly. Some implementation details and initial computational experience are presented.

Keywords: Integer Programming, Coloring, column generation, multi-coloring.

1. INTRODUCTION

The graph multi-coloring problem is a generalization of the well-known graph coloring problem. Given a graph, the (node) coloring problem is to assign a single color to each node such that the colors on adjacent nodes are different. For the multi-coloring problem, each node must be assigned a preset number of colors and no two adjacent nodes may have any colors in common. The objective is to accomplish this using the fewest possible number of colors.

Like the graph coloring problem, the multi-coloring problem can model a number of applications. It is used in scheduling ([7]) where each node represents a job, edges represent jobs that cannot be done simultaneously, and the colors represent time units. Each job requires multiple time units (the required number of colors at the node), and can be scheduled preemptively. The minimum number of colors then represents the makespan of the instance. Multi-colorings also arise in telecommunication channel assignment where the nodes represent transmitters, edges represent interference, and the transmitters send out signals on multiple wavelengths (the colors) [14]. It is due to this application in telecommunications that multi-coloring, as well as generalizations that further restrict feasible colorings, dates back to the 1960s. Aardal et al. ([1] provide an excellent survey on these problems.

The multi-coloring problem can be reduced to graph coloring by replacing each node by a clique of size equal to the required number of colors. Edges are then replaced with complete bipartite graphs between the corresponding cliques. Such a transformation both increases the size of the graph and embeds an unwanted symmetry into the problem. It is therefore useful to develop specialized algorithms that attack the multi-coloring problem directly.

Johnson, Mehrotra, and Trick [9] included the multi-coloring problem in a series of computational challenges, and provide a testbed of sample instances. Prestwich [17] developed a local search algorithm for this form of the multi-coloring problem and compared that approach to a satsifiability-based model. Without lower bounds or exact solutions to simple problems, however, it is difficult to evaluate these heuristic approaches.

We suggest an approach based on an integer programming formulation of the graph multi-coloring problem. This formulation, called the *independent set formulation*, has a variable for each independent set in the graph. In our previous work on graph coloring problems [12], we demonstrated that despite the enormous number of variables in this formulation, it is possible to develop an effective column generation technique for the coloring problem. We used appropriate branching rules and tested our branch-and-price approach on a variety of coloring instances. Encouraged by the effectiveness of such a method for coloring problems, we discuss the extension of such an approach on graph multi-coloring problems. This extension is independently interesting due particularly to the non-binary nature of the variables. Most examples of branch-and-price use binary variables, which results in now-routine branching rules. With non-binary variables, we need to explore new and intriguing approaches to branching.

In Section 2, we develop the independent set formulation of the graph multi-coloring problem and discuss various advantages of the formulation. In Section 3, we summarize the techniques for generating columns in this formulation and outline one method for such generation. In Section 4, we discuss the branching rules that are necessary to be developed for a full branch-and-price method. In Section 5, we describe some initial computational results and conclude with some directions for future exploration.

2. A COLUMN GENERATION MODEL

Let $G = (V, E)$ be an undirected graph on V, the set of vertices, with E being the set of of edges. Let $|V| = n$ and $|E| = m$. Let w_i be an integer weight associated with a node $i \in V$ giving the required number of colors at the node. When $w_i = 1$, for all $i \in V$, then the problem is the usual vertex coloring problem.

A *multi-coloring* of G is an assignment of w_i labels to each vertex i such that the endpoints of any edge do not have any common label. A *minimum multi-coloring* of G is a multi-coloring with the fewest different labels among all possible multi-colorings.

An *independent set*, S of G is a set of vertices $S \subseteq V$ such that there is no edge in E connecting any pair of nodes in S. Clearly in any coloring of G, all vertices with the same label comprise an independent set. A *maximal independent set* is an independent set that is not strictly included in any other independent set.

The problem of finding a minimum multi-coloring in a graph can be formulated in many ways. For instance, letting x_{ik}, $i \in V$, $1 \le k \le K$ be a binary variable that is 1 if label k is assigned to vertex i and 0 otherwise, where K represents an upper bound on the number of labels needed to obtain a valid multi-coloring of the graph, the problem can be formulated as follows:

$$
\begin{array}{llcll}
\textit{Minimize} & y & & & \\
\text{s.t.} & x_{ik} + x_{jk} & \le & 1 & \forall\,(i,j) \in E,\; k = 1,\dots,K \\
& \sum_k x_{ik} & = & w_i & \forall\, i \in V \\
& y & \ge & k x_{ik} & \forall\, i \in V,\; k = 1,\dots,K \\
& x_{ik} & \in & \{0,1\} & \forall\, i \in V,\; k = 1,\dots,K.
\end{array}
$$

We will refer to this formulation as (VC). While correct, (VC) is difficult to use in practice. One obvious problem is the size of the formulation. Since K can be quite large, the formulation can have up to nK variables and $2Km + n$ constraints. Given the need to enforce integrality, this formulation becomes computationally intractable for all except the smallest of instances. This is especially true because the linear programming relaxation is extremely fractional. To see this, note that even when all $w_i = 1$, the solution, $x_{ik} = 1/K$ for every (i, k) is feasible whenever $K \geq 2$.

A second, less obvious, problem involves the symmetry of the formulation. The variables for each k appear in exactly the same way. This means that it is difficult to enforce integrality in one variable without problems showing up in the other variables. This is because any solution to the linear relaxation has an exponential number (as a function of K) of representations. Therefore, branching to force x_{i1} to take on integral values does little good because it results in another representation of the same fractional solution in which x_{i2} takes on the old value of x_{i1} and vice-versa.

To address this problem, we consider a formulation with far fewer constraints that does not exhibit the same symmetry problems as our first formulation. Let T be the set of all maximal independent sets of G. We create a formulation with binary variables, x_t, for each $t \in T$. In this formulation, $x_t = k$ implies that independent set t will be given k unique labels, while $x_t = 0$ implies that the set does not require a label. The minimum multi-coloring problem is then the following (denoted (IS)):

$$
\begin{array}{llll}
\textit{Minimize} & \sum_{t \in T} x_t & & \\
\textit{Subject to} & \sum_{\{t: i \in T\}} x_t & \geq & w_i \quad \forall\, i \in V \\
 & x_t & \geq 0 \textit{ and integer} & \forall\, t \in T.
\end{array}
$$

This formulation can also be obtained from the first formulation by using a suitable decomposition scheme as explained in [10] in the context of general mixed integer programs. The formulation (IS) has only one constraint for each vertex, but can have a tremendous number of variables. Note that a feasible solution to (IS) may assign more than the specified number of labels to a vertex since we include only maximual independent sets in the formulation. This can be remedied by using any correct subset of the assigned multiple labels as the labels for the vertex. The alternative would be to allow non-maximal sets in T and to require equalities in (IS). In view of the ease of correcting the problem versus the great increase in problem size that would result from expanding T, we choose the given formulation.

This formulation exhibits much less symmetry than (VC): vertices are combined into independent sets and forcing a variable to 0 means that the vertcies comprising the corresponding independent set will not receive the same color in the solution. Furthermore, it is easy to show [10] that the bound provided by the linear relaxation of (IS) will be at least as good as the bound provided by the linear relaxation of (VC).

The fact remains, however, that (IS) can have far more variables than can be reasonably handled directly. We resolve this difficulty by using only a subset of the variables and generating more variables as needed. This technique, called *column generation*, is well known for linear programs and has emerged as a viable technique for a number of integer programming problems [5, 12]. The need to generate dual variables (which requires something like linear programming) while still enforcing integrality makes column generation procedures nontrivial for integer programs. The procedures need to be suitably developed and their effectiveness is usually dependent on cleverly exploiting the characteristics of the problem.

The following is a brief overview of the column generation technique in terms of (IS). Begin with a subset $\bar{T}$ of independent sets. Solve the linear relaxation (replace the integrality constraints on x_s with nonnegativity) of (IS) restricted to $t \in \bar{T}$. This gives a feasible solution to the linear relaxation of (IS) and a dual value π_i for each constraint in (IS). Now, determine if it would be useful to expand $\bar{T}$. This is done by solving the following *maximum weighted independent set* problem (MWIS):

$$\begin{array}{llcll} \textit{Maximize} & \sum_{i \in V} \pi_i z_i & & & \\ \textit{Subject to} & z_i + z_j & \leq & 1 & \forall\,(i,j) \in E \\ & z_i & \in & \{0,1\} & \forall\, i \in V. \end{array}$$

If the optimal solution to this problem is more than 1, then the z_i with value 1 correspond to an independent set that should be added to $\bar{T}$. If the optimal value is less than or equal to 1, then there exist no improving independent sets: solving the linear relaxation of (IS) over the current $\bar{T}$ is the same as solving it over T.

This process is repeated until there is no improving independent set. If the resulting solution to the linear relaxation of (IS) has x_t integer for all $t \in \bar{T}$, then that corresponds to an optimal solution to (IS) over T. When some of the x_t are not integer, however, we are faced with the problem of enforcing integrality.

To complete this algorithm, then, we need to do two things. First, since (MWIS) is itself a difficult problem, we must devise techniques to solve it that are sufficiently fast to be able to be used repeatedly. Second, we must find a way of enforcing integrality if the solution to the linear relaxation of (IS) contains fractional values. Standard techniques of enforcing integrality (cutting planes, fixing variables) make it difficult or impossible to generate improving independent sets. We discuss these two problems in the next two sections.

3. SOLVING THE MAXIMUM WEIGHTED INDEPENDENT SET PROBLEM

The maximum weighted independent set problem is a well-studied problem in graph theory and combinatorial optimization. Since a clique is an independent set in the complement of a graph, the literature on the maximum weighted clique is equally relevant. Various solution approaches have been tried, including implicit enumeration [6], integer programming with branch and bound [3, 4], and integer programming with cutting planes [2, 15]. In addition, a number of heuristics have been developed [16] and combined with general heuristic methods such as simulated annealing [8]. In this section, we outline a simple recursive algorithm based on the work of [11] and describe a simple greedy heuristic that can be used to reduce the need for the recursive algorithm.

The basic algorithm for finding a maximum weighted independent set (MWIS) in the graph $G(V, E)$ is based on the following insight. For any subgraph $G_1(V_1, E_1)$ of G, and a vertex $i \in V_1$, the MWIS in G_1 is either the MWIS in G_1 restricted to $V_1/\{i\}$ or it is i together with the MWIS in AN(i), where AN(i) is the *anti-neighbor* set of i: the set of all vertices j in V_1 such that $(i, j) \notin E_1$. This insight, first examined in [11] for the unweighted case, leads to the following recursion which can be turned into a full program:

$$\text{MWIS}(V_1 \cup \{k\}) = \max(\text{MWIS}(V_1), \text{MWIS}(\{k\} \cup \text{AN}(k))),$$

where MWIS(S) represents the maximum weighted independent set in the subgraph of G induced by the set of nodes in S.

While this approach is reasonably effective for graphs that are not too sparse, it can be improved by appropriately ordering the vertices to add to V_1. The following have been shown to be effective in reducing the computational burden of the recursion:

- Begin with V_1 equal to a heuristically found independent set. We use a simple greedy approach to find such a set, with the nodes ordered by node weight.
- Order the remaining vertices in order of degree from lowest to highest, and add them to V_1 in that order. During the final stages of the recursion, it is important to keep the anti–neighbor set small in order to solve the MWIS on as small a graph as possible. Since vertices with high degree have small anti–neighbor sets, those should be saved for the end.
- Use simple bounds to determine if a branch of the recursion can possibly return a MWIS better than the incumbent. For instance, if the total weight of the set examined is less than the incumbent, the incumbent is necessarily better, so it is unnecessary to continue the recursion.
- Use a faster code for smaller problems. It appears that a weighted version of the method of Carraghan and Pardalos [6] is faster for smaller problems. This is particularly the case since it is able to terminate when it is clear that no independent set is available that is better than the incumbent. In our tests, which use relatively small graphs, we use a variant of Carraghan and Pardalos for all except the first level of recursion, which echoes the results of Khoury and Pardalos in the unweighted case.

In the context of our column generation technique, it is not critical that we get the best (highest weight) maximal independent set: it is sufficient to get any set with weight over 1. This suggests that a heuristic approach for finding an improving column may suffice in many cases. It is only when it is necessary to prove that no set exists with weight over 1 (or when the heuristics fail) that it is necessary to resort to the recursion. There are many heuristics for weighted independent sets. The simplest is the greedy heuristic: begin with (one of) the highest weighted vertices and add vertices in nonincreasing order of their weight making certain that the resulting set remains an independent set.

This heuristic, in addition to being simple, is very fast, and seems to work reasonably well. The resulting independent set can either be added directly to (IS) (if it has value over 1) or can be used as a starting point for the recursion.

4. BRANCHING RULE

A difficult part about using column generation for integer programs is the development of branching rules to ensure integrality. Rules that are appropriate for integer programs where the entire set of columns is explicitly available do not work well with restricted integer programs where the columns are generated by implicit techniques.

The fact that the variables in (IS) are general integers, rather than binary variables, makes this issue even more difficult. For binary variables, the Ryan-Foster [18] branching rule is generally effective, but that rule cannot be used for general integer variables. For (single-color per node) graph coloring, given a solution to (IS), the Ryan-Foster rule identifies two nodes i and j, such that there is a fractional independent set that includes both i and j. The branching is then on whether i and j have the same color or different colors. For the purposes of generating improving independent sets, this involves either contracting two nodes into one or adding an edge to the graph, respectively, as developed in [12]. Such changes do not affect the operation of the MWIS algorithm.

For general integers, it is not necessarily the case that there will be a pair of vertices with a fractional number of colors in common. Vanderbeck [19] does show there are sets of nodes V_1 and V_2 such that the x values for all independent sets that contain all nodes in V_1 and no nodes in V_2 is fractional. If we let $S(V_1, V_2)$ represent the currently generated independent sets that contain all of V_1 and none of V_2, this leads to a branching rule with

$$\sum_{s \in S(V_1,V_2)} x_s \leq k$$

in one branch, and

$$\sum_{s \in S(V_1,V_2)} x_s \geq k+1$$

in the other. This can complicate the solving of the subproblem (MWIS) since either case involves adding a constraint to (IS). This constraint leads to a dual value that must be considered in the MWIS subproblem.

This problem can be addressed in one of two ways. Vanderbeck [19] gives an approach where multiple subproblems are solved without modifying the structure of the subproblem (in our case, MWIS). This approach has the advantage of keeping the subproblem algorithm the same, at the expense of requiring the solution of multiple subproblems. Further, this approach has the disadvantage that the branching rule needs to be more complicated than the node-pair rule given by the Ryan-Foster rule. Instead, the branching constraints need to consist of nested sets of constraints.

The alternative approach is to directly embed the dual values associated with branching constraints into the subproblem. To do this, we will have to modify the solution approach to MWIS to allow costs on arbitrary pairs of sets (V_1, V_2). This dual value is charged for any independent set that contains all of V_1 and none of V_2.

Fortunately, this is a straightforward modification of the implicit enumeration approach in [12], similar to the modification we proposed in the context of solving clustering problems [13] where the costs only appeared on edges between nodes.

The key aspect of our implicit enumeration is that, at each step, the nodes of the graph are divided into three sets: those that will be in the independent set (I), those that are definitely not in the independent set (NI), and those for which their status is unknown (UN). The duals associated with (V_1, V_2) can similarly be assigned one of three states: definitely to be charged (C), definitely not to be charged (NC) and "not yet determined" (UC). For instance, if the current independent set contains a member of V_2 we know that the corresponding dual on (V_1, V_2) will not be charged.

At each stage of the implicit enumeration, we can calculate an upper bound by adding in the duals for all nodes in I, all the positive duals in NI, all duals in C, and all positive duals in UC. The lower bound is the sum of the duals in I and C. We can strengthen the bounds somewhat by taking the dual for any entry in UC containing just one node in UN and moving that dual value to the UN node. This gives a valid recursion for the case of dual values on arbitrary node sets.

5. COMPUTATIONAL DETAILS

Our current implementation focuses on first optimizing the LP relaxation of (IS) via column generation. Then we determine the best integer solution to the restricted (IS) formulation comprising of the columns generated to optimize the LP relaxation at the root node of the branch-and-price tree. Here we provide some implementation details and initial computational results that we have obtained.

5.1 Implementation Issues

We generate a feasible initial multi-coloring using the greedy MWIS heuristic repeatedly until all nodes are colored at least once. This gives us an initial solution to the multi-coloring problem as well as a number of columns to add to our linear program. We then generate columns to improve the linear program. The following discussion pertains to generation of columns to improve the linear program.

Improving the Linear Program.

Improving Column. As mentioned earlier, any solution to the MWIS with value greater than 1 represents an improving column for the linear program. In our current implementation, we set a target to 3.0 and our MWIS algorithm either returns the first such solution it finds, failing which, it finds the exact solution. We have also experimented with changing this target value to a higher number initially (an approach to find a *good* set of columns as fast as possible) and then decreasing its value later on in the column generation. The effort required to solve some difficult problems can be substantially reduced by suitably altering this target value.

Ordering the Nodes. The order in which the nodes are to be considered can be specified in our MWIS algorithm. We have found that ordering the nodes independently by nonincreasing weights or by nonincreasing degree is not as efficient as ordering them by considering both at the same time. In our experiments we order the nodes in nonincreasing values of the square root of the degree of the node times the weight of the node.

Column Management. Another approach to optimizing the linear program more quickly is to generate several columns rather than a single column [5] at every iteration. For example, one could use improvement algorithms that take existing columns with reduced cost equal to zero and try to construct columns that might improve the linear program. In our experiments, we generated more candidates by determining other independent sets at each iteration such that every node belonged to at least one indpendent set being added.

5.2 Computational Results

In our computational experiments, we use instances drawn from a large number of sources. Our goal is to determine the robustness of the approach. For some of these graphs, the coloring problem has no real interpretation. We use these graphs as examples of structured graphs, rather than just experimenting on random graphs. These graphs come from a large test set at http://$mat.tepper.cmu.edu/COLOR$04.

Currently, we have not implemented the branching scheme. Rather, we use the standardized branching to determine an integer solution from among the independent sets generated at the root node to optimize the corresponding LP relaxation of the (IS) formulation. Hence our current implementation provides an optimization-based heuristic procedure. We report our results in Tables 1 and 2. The instance name identifies the problem from the test set. The objective values corresponding to the optimal LP relaxation solution and the intger solution obtained by our method are listed under the columns labeled LP, and Heur, respectively. The gap between these two objective values and the computational time in seconds to optimize the linear relaxation and then to determine the integer solution are listed in the next three columns. The column labeled *cons* lists the number of constraints in the corresponding (IS) equal to the number of vertices in the graph. The number of independent sets generated to optimize the LP relaxation is listed under the column labeled *vars*. The computational results reported here are limited to the best integer solution found in at most 1000 seconds using CPLEX default branching scheme on DEC ALPHA workstation. As can be seen from the gap between the LP bound and the corresponding (heuristic) integral solution obtained by our methodology, this branch-and-price framework looks promising for finding optimal multi-coloring solutions for small to moderate size graphs. In Table 1, we report results on geometric graphs with up to 120 nodes. The best integer solution found for these is within 1 of the optimal multi-coloring in the worst case. The cpu time is also reasonable. A similar performance is seen for the random graphs of up to 100 nodes except for R100-1ga where the gap is 2 between the LP bound and the best integer solution found in 1000 seconds. The gaps are higher for some miscellaneous graphs in Table 2.

5.3 Further Research

A full implementation of the branching is necessary to complete the branch-and-price framework proposed here. Based on the initial results, there is hope that the LP bound is strong and one may not need to have a very deep branch-and-price tree to find optimal multi-colorings for many structured graphs. Further exploration will explore the robustness of this framework for general graphs.

It will also be interesting to see the comparison between using this branch-and-price scheme with a branch-and-price scheme that uses modified branching scheme proposed by Vanderbeck [19].

Finally, it will be interesting to see if this framework can be suitably exploited to solve other variations and extensions of coloring problems.

Table 1. Results for Geometric Graphs

Instance	LP	Heur	Gap	cpu-lp	cpu-ip	cons	vars
geom20	28.00	28	0	0	0	20	31
geom20a	30.00	30	0	0	0	20	29
geom20b	8.00	8	0	0	0	20	34
geom30	26.00	26	0	0	0	30	49
geom30a	40.00	40	0	0	0	30	65
geom30b	11.00	11	0	0	0	30	67
geom40	31.00	31	0	0	0	40	76
geom40a	46.00	46	0	0	0	40	69
geom40b	14.00	14	0	0	0	40	96
geom50	35.00	35	0	0	0	50	96
geom50a	61.00	61	0	0	0	50	106
geom50b	17.00	18	1	0	0	50	121
geom60	36.00	36	0	0	0	60	124
geom60a	65.00	65	0	0	0	60	120
geom60b	22.00	22	0	0	0	60	129
geom70	44.00	44	0	0	0	70	131
geom70a	71.00	71	0	0	0	70	130
geom70b	22.00	23	1	0	1	70	160
geom80	63.00	63	0	0	0	80	130
geom80a	68.00	68	0	0	0	80	168
geom80b	25.00	26	1	0	1	80	211
geom90	51.00	52	1	0	1	90	171
geom90a	65.00	66	1	0	5	90	243
geom90b	28.00	29	1	0	2	90	213
geom100	60.00	60	0	0	1	100	180
geom100a	81.00	81	0	0	2	100	241
geom100b	30.00	31	1	0	25	100	276
geom110	62.00	63	1	0	10	110	212
geom110a	91.00	92	1	0	141	110	260
geom110b	37.00	37	0	0	1	110	214
geom120	63.50	64	0	2	167	120	268
geom120a	93.00	94	1	2	303	120	329
geom120b	34.00	35	1	1	462	120	302

Table 2. Results for Random and Other Miscellaneous Graphs

Instance	LP	Heur	Gap	cpu-lp	cpu-ip	cons	vars
R50-1ga	12.00	12	0	0	0	50	91
R50-1gba	45.00	45	0	0	0	50	82
R50-5ga	28.12	29	0	0	0	50	482
R50-5gba	99.68	100	0	0	0	50	441
R50-9ga	64.00	64	0	0	0	50	253
R50-9gba	228.00	228	0	0	0	50	177
R75-1ga	14.00	15	1	1	1	70	262
R75-1gba	53.00	54	1	0	4	70	224
R75-5ga	37.17	38	0	2	2	75	1290
R75-5gba	130.84	131	0	2	4	75	1262
R75-9ga	93.50	94	0	0	0	75	354
R75-9gba	328.00	328	0	0	0	75	372
R100-1ga	15.00	17*	2	10	1000	100	612
R100-1gba	56.00	57	1	4	127	100	492
R100-5ga	41.96	43	1	7	38	100	2292
R100-5gba	152.57	153	0	6	228	100	2171
R100-9ga	117.29	118	0	0	0	100	786
R100-9gba	421.50	422	0	0	0	100	640
myciel3	10.50	11	0	0	0	11	27
myciel3b	31.50	32	0	0	0	11	24
myciel4	11.71	12	0	0	0	23	83
myciel4b	38.80	39	0	0	0	23	70
myciel5	13.32	14	0	0	0	47	243
myciel5b	44.83	45	0	0	0	47	189
myciel6	15.47	16	0	1	3	95	578
myciel6b	57.14	58	0	2	1	95	595
myciel7	16.37	17	0	30	3	191	1379
myciel7b	60.74	61	0	18	23	191	1096
queen8-8	28.00	29	1	0	1	64	266
queen8-8b	113.00	113	0	0	0	64	148
queen9-9	35.00	36	1	0	2	81	215
queen9-9b	135.00	135	0	0	2	81	242
queen10-10	38.00	40	2	1	123	100	291
queen10-10b	136.00	136	0	0	42	100	282
queen11-11	41.00	44*	3	1	1000	121	349
queen11-11b	140.00	142*	2	3	1000	121	443
queen12-12	42.00	47*	5	38	1001	144	701
queen12-12b	163.0	165.0*	2	1	1000	144	376
DSJC125.1	19.00	21	2	2	57	125	321
DSJC125.1b	67.00	68	1	2	63	125	368
DSJC125.5	52.87	55*	2	20	1001	125	3591
DSJC125.5b	161.5	164.0*	2	19	1000	125	3733
DSJC125.9	139.00	140	1	1	1	125	1388
DSJC125.9b	496.25	497	0	1	0	125	1270

References

[1] Aardal, K.I., S.P.M. Van Hoesel, A.M.C.A. Koster, C. Manino, and A. Sassano. (2001). Models and Solution Techniques for Frequency Assignment Problems 4OR 1:4, 261–317.

[2] Balas, E. and H. Samuelsson. (1977). A node covering algorithm, *Naval Research Logistics Quarterly 24:2*, 213–233.

[3] Balas, E. and J. Xue. (1991). Minimum weighted coloring of triangulated graphs, with application to maximum weight vertex packing and clique finding in arbitrary graphs, *SIAM Journal on Computing 20:2*, 209–221.

[4] Balas, E. and C. S. Yu. (1986). Finding a maximum clique in an arbitrary graph, *SIAM Journal on Computing 15:4*, 1054–1068.

[5] Barnhart, C., E. L. Johnson, G. L. Nemhauser, M. W. P. Savelsbergh, and P. H. Vance. (1998). Branch-and-Price: Column Generation for Huge Integer Programs, *Operations Research 46:3*, 316–329.

[6] Carraghan, C. and P. M. Pardalos. (1990). An exact algorithm for the maximum clique problem, *Operations Research Letters 9*, 375–382.

[7] Coffman Jr., E.G., M.R. Garey, D.S. Johnson, and A.S. Lapaugh. (1985). Scheduling File Transfers *SIAM Journal on Computing 14:4*, 743-780.

[8] Jerrum, M. (1992). Large cliques elude the metropolis process, *Random Structures and Algorithms 3:4*, 347–360.

[9] Johnson, D.S. A. Mehrotra, and M.A. Trick. (2002). Computational Challenge on Graph Coloring and its Generalizations International Symposium on Mathematical Programming, Copenhagen, Denmark.

[10] Johnson, E.L. (1989). Modeling and strong linear programs for mixed integer programming, *Algorithms and Model Formulations in Mathematical Programming, NATO ASI 51*, S.W. Wallace (ed.), Springler-Verlag Berlin, Heidelberg, 1-43.

[11] Khoury, B.N. and P. M. Pardalos. (1996). An algorithm for finding the maximum clique on an arbitrary graph, *Second DIMACS Challenge: Cliques, Coloring, and Satisfiability, DIMACS Series on Discrete Mathematics and Theoritical Computer Science*, D. S. Johnson and M. A. Trick (eds.), American Mathematical Society, Providence.

[12] Mehrotra, A. and M. A. Trick. (1996). A column generation approach for exact graph coloring, *INFORMS Journal on Computing*, **8:4**, 133-151.

[13] Mehrotra, A. and M. A. Trick. (1998). Cliques and Clustering: A Combinatorial Approach, *Operations Research Letters*, **22:1**, 1-12.

[14] Narayanan, L. (2002). Channel Assignment and Graph Multi-coloring, in *Handbook of Wireless Networks and Mobile Computing*, Wiley.

[15] Nemhauser, G.L. and L. E. Trotter. (1975). Vertex packings: Structural properties and algorithms, *Mathematical Programming 8*, 232–248.

[16] Pittel, B. (1982). On the probable behaviour of some algorithms for finding the stability number of a graph, *Mathematical Proceedings of the Cambridge Philosophical Society 92*, 511–526.

[17] Prestwich, S. (2006). Generalized Graph Coloring by a Hybrid of Local Search and Constraint Programming, *Discrete Applied Mathematics*, to appear.

[18] Ryan, D.M. and B.A. Foster. (1981). An integer programming approach to scheduling, in *Computer Scheduling of Public Transport Urban Passenger Vehicle and Crew Scheduling*, North-Holland, Amsterdam, 269-280.

[19] Vanderbeck, F. (2005). Branching in Branch-and-Price: A Generic Scheme, manuscript, Applied Mathematics, University Bordeaux 1, F-33405 Talence Cedex, France.

A GENETIC ALGORITHM FOR SOLVING THE EUCLIDEAN NON-UNIFORM STEINER TREE PROBLEM

Ian Frommer[1] and Bruce Golden[2]

[1]*Department of Mathematics, United States Coast Guard Academy, New London, CT 06320, ifrommer@exmail.uscga.edu*

[2]*R.H. Smith School of Business, University of Maryland, College Park, MD 20742, BGolden@rhsmith.umd.edu*

Abstract In this paper, we present a genetic algorithm developed to solve a variation of the Euclidean Steiner tree problem. In the problem we consider, the non-uniform Steiner tree problem, the goal is to connect a set of nodes located on an uneven landscape, in a tree in the cheapest possible way. Using a combination of novel operators, our genetic algorithm is able to find optimal or near-optimal solutions in a small fraction of the time taken by an exact algorithm. In tutorial fashion, we describe the problem and some of its applications, our algorithm along with other possible approaches, and present images of several solved problem instances.

Keywords: Steiner tree, non-uniform, genetic algorithm.

1. INTRODUCTION

We consider a variation of a well-known network optimization problem that has applications in the design of transmission networks. The Euclidean Non-Uniform Steiner tree problem (ENSTP) we consider is based on the minimal spanning tree problem (MSTP). The goal in the MSTP is to connect a set of nodes in a minimal cost tree. In a Steiner tree problem, a set of nodes must be connected, but additional nodes may be used in the tree to help reduce the total cost. The additional nodes that are used, if any, are known as *Steiner nodes*. In the ENSTP, all of the nodes are situated on a plane in which costs are associated with each location. The costs of the edges connecting the nodes depend on the costs of the locations through which the edges pass. This problem is relevant to situations in which a network must be constructed on a non-uniform cost landscape. For example, consider the problem of taking a set of locations in a city and connecting them via an underground cable-based communications network. If laying cable underground requires digging up the street, then it will usually be more costly to do so in a downtown location than in an industrial section on the edge of town. Other applications include transmission lines, printed circuits, heating and ventilation, water supply, fire sprinkler, and drainage systems [1–3]. In this paper, we present a genetic algorithm (GA) to solve this problem. We show that the GA finds optimal or near-optimal solutions to a variety of test problems and in a small fraction of the time required by an exact solver.

There are a number of variants of the Steiner tree problem. (Surveys of some of the standard versions may be found in [1, 4–7].) In the Euclidean version, the nodes are points in the Euclidean plane and the edge costs are the Euclidean distances. In the Rectilinear Steiner tree problem, edges must be oriented horizontally or vertically. In this case, the edge costs are the so-called *taxicab* distances. A modified version of the Rectilinear problem considers nodes located on a hexagonal grid [8, 9]. One can also study the problem in an arbitrary network (i.e., a weighted graph). Later, we discuss how we convert our problem to one on a network in order to find optimal solutions for small to medium-sized problem instances.

A problem related to the ENSTP is the Steiner tree problem with obstacles (see [4]). As a practical example, consider configuring the layout of an electrical system on a building floor in the presence of walls and columns that may not be penetrated. This can be viewed as a specific case of the ENSTP in which the cost structure is uniform except for the locations of obstacles, where the costs are infinite. Zachariasen and Winter [10] present an exact algorithm for the Euclidean Steiner tree problem with polygonal obstacles.

In VLSI micro-chip design, numerous components on a chip need to be connected efficiently in a tree with wires. This can be achieved by solving a Rectilinear Steiner tree problem. To anticipate requirements of later stages of the chip design process, it may be necessary for the Steiner tree to avoid certain locations. Thus, the problem can be viewed as a Rectilinear Steiner tree problem with obstacles. Alpert et al. [3] develop an algorithm to solve this problem, known as the buffered Steiner tree construction problem.

In some applications, Steiner nodes corresponding to bends or junctions may require the use of fixtures or other hardware that can increase the overall cost of the tree [2]. Charging additional penalties for Steiner nodes can reflect these issues. This *node-weighted* Steiner tree problem is described in [4]. In earlier work [11], we considered this problem within the context of the ENSTP.

The Steiner tree problem can also be studied in three dimensions. With applications involving 3-D VLSI layout in mind, Kanemoto et al. [12] solves the rectilinear 3-D problem using a genetic algorithm. Stanton and Smith [13] use 3-D Steiner trees to model minimum energy configurations of large complex molecules, such as proteins and DNA.

Since in general, Steiner tree problems have been shown to be NP-hard [14, 15], algorithms designed to find optimal solutions have exponential computational complexity in the worst case. This motivates our development of an approximation algorithm. Heuristic approaches to find near-optimal solutions have been proposed for several other variants of the Steiner tree problem. Gröpl et al. [16] review several mostly greedy algorithms that have been applied to the Steiner tree problem in networks. Numerous randomized algorithms [17–21] exist as well, some of which use GAs.

For the ENSTP, Coulston [9] applies a GA in which solutions are encoded using full components, a kind of sub-tree. Coulston defines an edge to be any path between two nodes, straight or otherwise. In our work, edges must be straight-lines, a restriction relevant to applications such as the design of circuits in which components are generally connected by straight-line segments of wire. Coulston utilizes a ring-growing algorithm, akin to a cellular automaton, to find minimal spanning trees on the set of terminal and Steiner nodes. His algorithm is applied to test cases with either uniform or random cost structures. In the latter case, the algorithm's Steiner tree costs are compared with MST costs. In contrast, we apply our algorithm to more structured landscapes including hills and pits, and compare the costs of the Steiner trees our algorithm finds with the optimal solutions. Owing to the complexity of his approach, and the difference in edge definition, we do not implement Coulston's algorithm.

In [11], we presented an early version of a GA to solve the ENSTP, applied it to some test cases and compared results to another fairly simple deterministic algorithm we developed. We also applied the algorithm to the node-weighted version of the ENSTP in which additional fees are charged for each Steiner node. This paper presents a more sophisticated GA that is able to find better solutions faster than the GA in [11]. In addition, we show that the GA compares quite favorably to an exact solver in terms of solution cost and run time.

This paper is organized as follows: In the next section, we formulate the problem, in Sect. 3, we describe our genetic-algorithm-based procedure, in Sect. 4, we present results, followed by conclusions including potential directions for future work.

2. PROBLEM FORMULATION

We formulate the problem in the same manner as in [11]. The goal is to connect a given set of nodes (terminal nodes) that are located on a surface in which each location has an associated cost. As in the usual Steiner tree problem, additional nodes not in the terminal-node set may be used to help connect the terminal nodes. Following [9], we situate the problem in a hexagonal grid. We fix the sizes of the grid cells throughout this paper, and so do not address the impact of grid cell size on the quality of solutions. Each hexagonal cell may

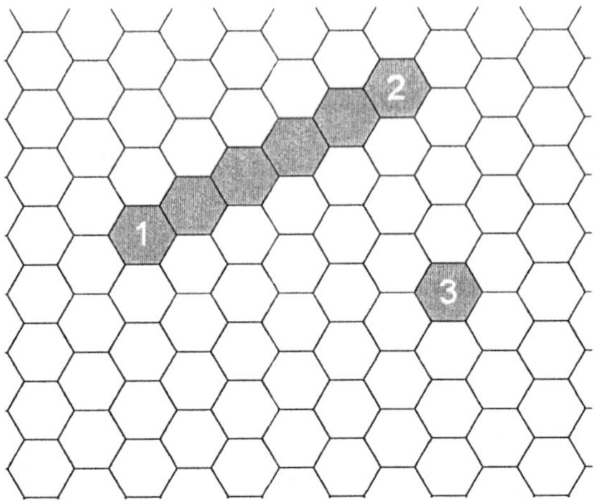

Figure 1. Hexagonal tiling of a 2-dimensional space. An edge is defined to be a straight line of cells. Nodes in cells 1 and 2 can be connected, but a node in cell 3 cannot be directly connected to either of the other two.

contain at most one node and has a cost associated with it. Nodes are assumed to be located at the center of their hex cells. Two nodes can be connected directly only if a straight line of cells can be drawn between the cells containing the two nodes (see Fig. 1). The cost of an edge connecting two nodes is equal to the sum of the costs associated with all the intermediate cells plus one half the costs of the two nodes' cells.

2.1 Network Steiner Tree Formulation

The use of the hexagonal grid and the restriction of one node per hexagonal cell simplify the search for optimal solutions because they reduce an otherwise uncountable search space to one of a finite size. In effect, this allows us to reduce the ENSTP to a Steiner tree problem in a network. This is illustrated in Fig. 2. The points in each cell represent potential terminal or Steiner nodes. Edges connect nodes to each of their neighbors forming a *triangle graph*. The cost of these nearest-neighbor edges is computed by adding together half of the cost of each cell the edge traverses. (Recall that we assume each node is located in the center of its cell, and so any edge incident to it traverses half of the cell.) The problem is now represented in network form. This enables us to find optimal solutions using an exact algorithm for the network problem.

Note that with the problem represented in this form, a GA is only one of many possible approaches. Numerous high performing algorithms for the Steiner tree problem in networks exist including preprocessors [22], approximate solvers [23], and exact solvers [24, 25]. Given access to one of these implementations, one could solve the problem with it and/or compare our GA to it. We use a GA to demonstrate a nice application of the GA to a combinatorial optimization problem and because it more easily allows for generalizations of the problem. For example, the node-weighted problem that we applied an earlier version of this GA to in [11] cannot be easily expressed as a Steiner tree problem in a network. Hence the high performing algorithms mentioned above would not apply in their current form.

In this paper, we solve small and medium-sized problems to optimality using the Dreyfus-Wagner algorithm [26]. This algorithm is not the fastest exact algorithm, but it is readily available (see [27] for a lucid presentation.) The exponential computational complexity of the Dreyfus-Wagner algorithm makes it impractical to solve larger problems. We will show that our GA compares favorably with the exact algorithm with respect to quality of solution. We consider an acceptable GA to be one whose mean plus one standard deviation is less than 4% above the value of the optimal solution on all test problems. We do not preprocess any of the instances. This could clearly benefit both algorithms, and should be undertaken in any future work.

Another benefit of representing the problem in a network is that it facilitates the determination of *shortest path minimal spanning trees* that we use to seed the initial population of the GA and to improve existing solutions. This is described in Sect. 3.

2.2 Problem Instances

The problem instances used in this paper are listed in Table 1; solutions to some are shown in Figures 8 through 11. (A full listing of the problem instance data and solutions can be found in [28].) They consist of an assortment of terminal node sets and grid cost structures. For terminal node sets, we used both randomly generated ones, and ones with specific structures, such as a ring. Some of the grid cost structures we used were random while others contained various combinations of hills and pits. While our sample problems are simpler than real-world examples, they test the algorithm on many of the basic topographic elements to be found in real world examples - hills and depressions, random layouts, etc. It is also possible to combine solutions from smaller, simpler problems such as these to solve larger, more complex problems.

3. THE GENETIC ALGORITHM

We experimented with many different GAs using a variety of operators and heuristic search strategies, such as population diversity and divide and conquer. Some were similar to ones used to solve the Euclidean Steiner tree problem. In

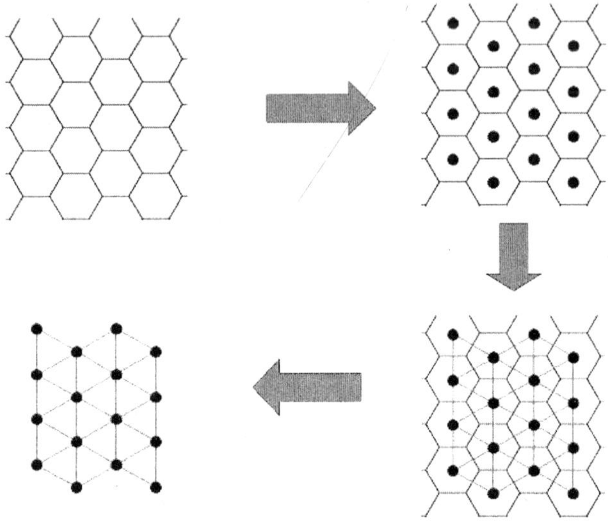

Figure 2. Transforming the ENSTP to a Steiner tree problem in a network. Starting with the hexagonal grid in the upper left, we designate potential node locations as in the upper right part of the figure. Next, each node location is connected to its nearest neighbors as the edges drawn in the lower right indicate. The cost of these edges is computed by adding together half of the cost of each cell the edge traverses. We now have a network and the grid can be discarded as in the lower left.

this paper, we describe the most successful GA, which was a more sophisticated version of the GA we developed in [11].

The GA is depicted schematically in Fig. 3. The solution encoding, fitness definition and the parent selection, crossover, and mutation operators are similar to those in our earlier GA. Though we review them briefly here, more information can be found in [11]. The most significant change with respect to the earlier algorithm is the method for generating the initial population.

As is often the case for GAs, our GA uses a large number of parameters such as population size, crossover and mutation probabilities, number of iterations, etc. Following common practice, we determined the parameter values using a combination of rules of thumb (see [29], for example) and trial and error. Since our goal is to engineer efficient networks, we are less concerned with the compactness of the algorithm than the quality of the solutions it produces.

Table 1. Problem Instances

Problem #	Grid Size	Grid Type	Terminal Node Set # of Nodes	Terminal Node Set Type
1	21 x 17	Hill	10	Random
2	21 x 17	Hill	7	Ring
3	21 x 17	2-Hill	10	Random
4	21 x 17	2-Hill	7	Ring
5	24 x 15	Random	7	Random
6	35 x 35	4-Hill	15	Random
7	35 x 35	Hill	15	Random
8	35 x 35	4-Hill	15	Ring
9	35 x 35	Hill	15	Ring
10	35 x 35	Random	15	Random
11	50 x 50	Hill	20	Random
12	50 x 50	2-Pit	20	Ring
13	80 x 80	Hill	32	Random

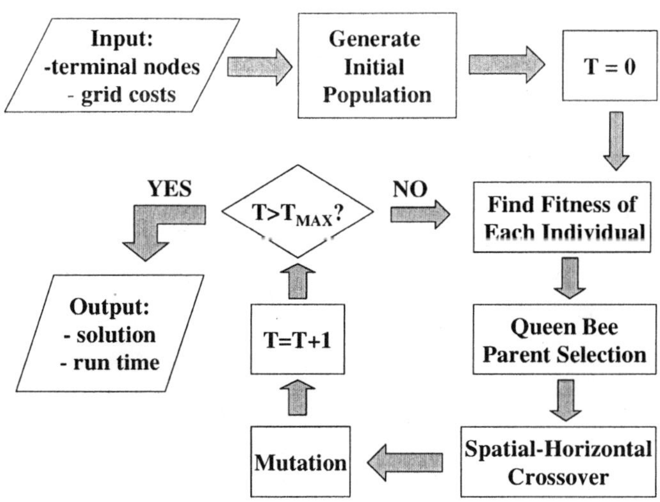

Figure 3. Flow chart for the genetic algorithm. For clarity, some details have been omitted.

3.1 Encoding and Initial Population

Each individual in the GA population is represented by a list of potential Steiner nodes to be used in a Steiner tree. The lists hold up to $4N$ node locations where N is the number of terminal nodes. In order to speed up the overall algorithm, we use two methods to seed the initial population with individuals that are better than purely random solutions. The first method is to generate what we call *shortest path minimal spanning trees*. Recall from Sect. 1 that we restrict edges to be straight line segments (see Fig. 4). To generate the seed solutions, we begin by temporarily redefining an edge to be the shortest path between its endpoint nodes. Such edges can have multiple turns in them. These edges can be found for all of the terminal nodes at once using an all-pairs shortest paths algorithm such as Floyd's algorithm [30]. The terminal nodes and this set of edges form a complete graph, upon which it is simple to find a minimal spanning tree using Prim's algorithm [31]. The resulting tree is what we refer to as the *shortest path minimal spanning tree* (see Fig. 5).

Because we are restricting the edges to be straight-line segments, we can convert the shortest path minimal spanning tree into a viable (by our edge definition) Steiner tree by introducing Steiner nodes at all turn and junction locations within the edges. This is demonstrated in Fig. 6. This procedure will yield one individual in the population by encoding the Steiner node locations that are introduced. We generate 10 such solutions by repeatedly perturbing the underlying grid cost structure by a small random amount. (We add a value taken from a uniform distribution ranging between 0 and 2.5 to the cost of each cell location. The unperturbed cell costs are between 0 and 1. The value 2.5 was chosen because it was found to be large enough to produce diverse solutions.)

Another 20 members of the initial population contain Steiner node locations that are randomly generated but with a bias towards low-cost regions of the grid. The remaining 10 members of the initial population have purely random Steiner node locations.

3.2 Fitness

Starting with the potential Steiner nodes of a given individual, the complete graph over those nodes and the terminal nodes is formed. We next find the MST and then prune away degree-1 Steiner nodes and their incident edges. The degree of a node is defined as the number of edges incident to that node. Any degree-1 Steiner node in a Steiner tree can be removed since it is not needed to connect the terminal nodes. Removing these nodes will lead to an improvement in the solution value, and may also cause the degree of other Steiner nodes to drop to one. This procedure is applied iteratively until all Steiner nodes have degree greater than one. The fitness of the individual is the cost of the resulting Steiner tree. This method for finding a Steiner tree

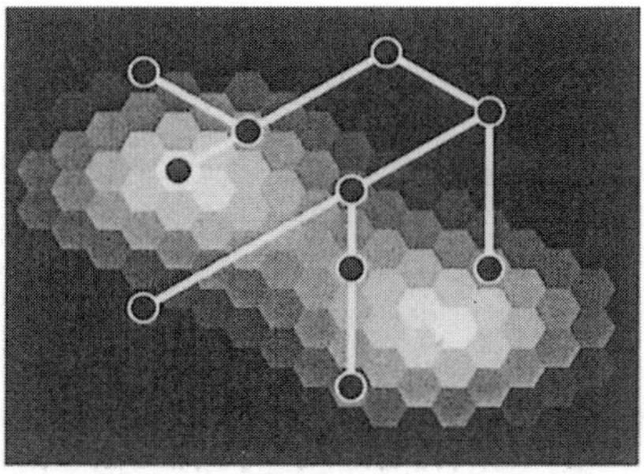

Figure 4. A minimal spanning tree (MST) using our straight-line edge definition. Black circles indicate terminal nodes. The lighter the cell, the more costly it is.

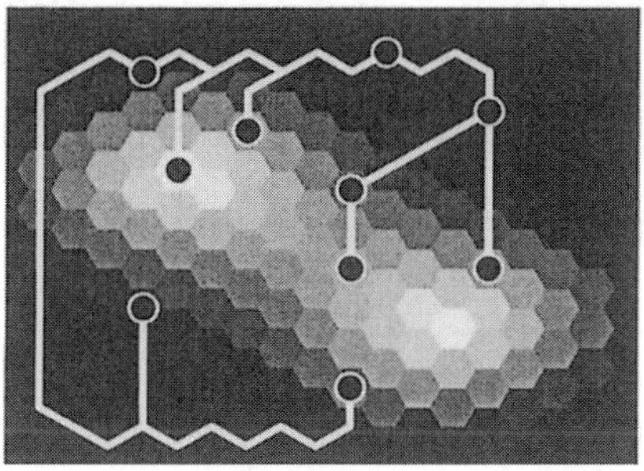

Figure 5. A *shortest path* MST for which an edge is taken to be the shortest path between the nodes it connects.

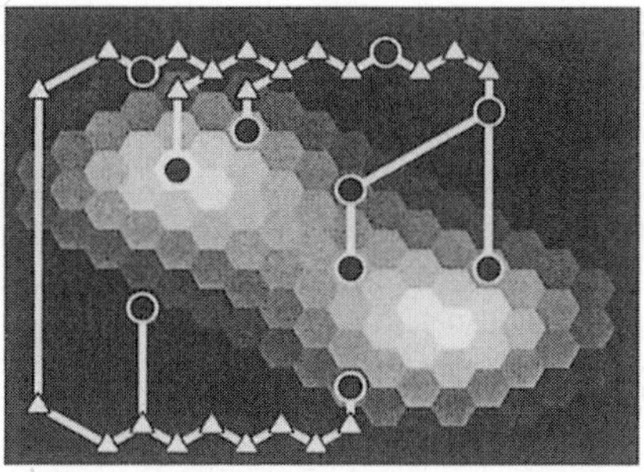

Figure 6. The MST in the previous figure can be viewed as a Steiner tree consistent with our edge definition if Steiner nodes (indicated by triangles) are designated at all bends and junctions. The initial population of our GA is seeded with individuals whose Steiner nodes are generated this way.

is quick, though not necessarily optimal for the individual's set of potential Steiner nodes.

3.3 Parent Selection and Crossover

The fittest individual, dubbed the Queen Bee, mates with each of the other individuals in the population [32]. This produces two offspring per pair of parents. The 40 fittest individuals out of the parents and offspring are chosen to replace the current population. The mating procedure, spatial-horizontal crossover, involves combining the potential Steiner nodes from Parent 1 located in the upper part of the grid with the potential Steiner nodes from Parent 2 located in the lower part of the grid, and vice versa. The grid is split into an upper and lower part at its vertical midpoint plus a normal random variable (of mean 0 and positive standard deviation). See Fig. 7.

Queen Bee selection appears to be successful for two reasons. First, it allows for the incremental improvement of an already good solution. It ensures that many of the best solution elements are preserved in each offspring. Second, unlike a more globally elitist scheme such as tournament or roulette wheel selection, it allows even the worst individuals to pass parts of their solutions to the next generation. This is advantageous since, as we learned when studying individual solutions, very poor ones often have high quality sub-trees within them.

The idea behind using crossover with a spatial component is that even though the problem can be viewed as a network problem, the nodes actually do have physical locations. This operator makes use of this spatial information by preserving some physically-near components of the tree while breaking up others. How the grid is split (horizontally, vertically, diagonally, etc.) is arbitrary. Since all individuals are used as parents for crossover, the probability of crossover is 1.

3.4 Mutation

If there are any intersecting edges in the Steiner tree, the solution value might be improved by introducing a new Steiner node at the intersection location. The solution improvement is possible because one of the four edges originating at that intersection can be deleted from the tree if a Steiner node is created there. This is implemented in the GA as a mutation that is applied to all individuals (i.e., it occurs with probability 1). The new Steiner nodes replace unused Steiner nodes, of which there are typically many, within the individual's chromosome. We also apply a second form of mutation that moves a randomly selected Steiner node in an individual to a random location. This occurs with probability 0.20 for all individuals except for the two fittest, which are exempt. This mutation is intended to keep solutions from stagnating at local minima. The probabilities we

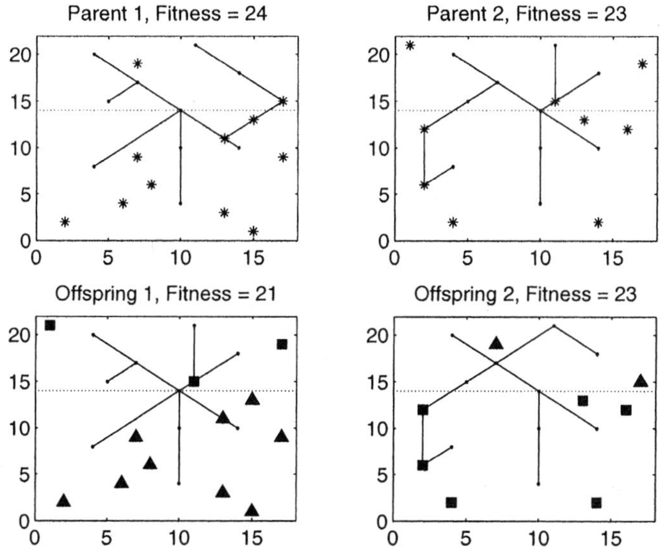

Figure 7. Horizontal crossover. The small dots in each plot represent the terminal nodes. In the top two plots, stars represent the Steiner nodes for the two parents. In the bottom two plots, triangles and squares represent the Steiner nodes in each offspring that came from Parents 1 and 2, respectively. The horizontal crossover location is indicated by the dotted line. The solid line segments indicate the Steiner tree edges.

use for crossover and the two mutations, along with the number of generations we run (70) were arrived at through trial and error.

After the algorithm completes a run, the coordinates of the Steiner nodes in the best final Steiner tree are reported along with run time. In addition, the program displays an image of the best final Steiner tree over a color or gray-scale grid cost structure, with terminal and Steiner nodes indicated (e.g., see Figures 8 through 11).

4. RESULTS

We ran the exact algorithm, our GA, and a standard GA on the first 10 problems in Table 1. We also ran our GA and the standard GA on problems 11, 12, and 13 in Table 1. Due to the prohibitively long running time required by the exact algorithm, we did not solve these last three problems to optimality. The idea behind comparing to a standard GA is to show that there is an advantage gained by the specific components we used (Queen Bee selection, spatial crossover, shortest path minimal spanning tree seeds, etc.). We are interested in comparing an intelligent algorithm that makes use of information from the

Table 2. Final Steiner Tree Costs for Each Algorithm. The values for each GA are calculated using 10 runs per problem. The numbers in parenthesis are the percent above the optimal value. The exact algorithm was not run on the last three problems because of their large size.

		Our GA			*Standard GA*		
#	*Optimal*	*best*	*mean*	*stdev*	*best*	*mean*	*stdev*
1	11.138	11.138(.00)	11.155(.15)	.01	11.215(.69)	11.360(1.99)	.10
2	10.001	10.001(.00)	10.012(.12)	.03	10.277(2.76)	10.416(4.15)	.08
3	4.806	4.806(.00)	4.806(.00)	.00	4.907(2.09)	4.942(2.84)	.04
4	4.158	4.158(.00)	4.158(.00)	.00	4.367(5.04)	4.562(9.73)	.14
5	5.605	5.605(.00)	5.605(.00)	.00	5.605(.00)	5.605(.00)	.00
6	26.606	26.653(.18)	27.104(1.87)	.21	30.015(12.8)	30.930(16.3)	.48
7	31.310	31.405(.30)	31.513(.65)	.07	34.016(8.64)	34.892(11.4)	.53
8	31.965	32.324(1.12)	32.574(1.91)	.19	38.450(20.3)	39.322(23.0)	.50
9	32.744	32.744(.00)	32.752(.02)	.02	33.381(1.95)	34.455(5.22)	.66
10	19.435	19.494(.31)	19.730(1.52)	.26	23.030(18.5)	23.760(22.3)	.46
11	—	32.227	32.479	.24	33.776	34.579	.44
12	—	38.234	38.240	.01	38.407	38.541	.08
13	—	72.325	72.471	.22	77.546	79.224	1.34

problem with an "off-the-shelf" GA that has no specific problem knowledge. Is this problem one in which a simple randomized algorithm can obtain good solutions? Or is it more difficult, requiring more intelligence on the part of the algorithm?

The standard GA uses tournament parent selection rather than Queen Bee selection, and one-point crossover instead of spatial-horizontal crossover. (In tournament selection, we draw subsets of size 5 from the population and choose the fittest 2 individuals in each subset to be parents. This process is repeated until 20 pairs of parents have been chosen. In one-point crossover, the parents are viewed as lists of Steiner node locations, and a position in the list is chosen at random. The entries before this position in parent 1 are combined with the entries after this position from parent 2 to create offspring 1, and vice versa to create offspring 2. The 40 offspring that are generated become the new population.) The standard GA uses a purely random initial population.

For each problem, both GAs were run 10 times at 70 iterations per run. The cheapest tree cost found by each run was returned and then the mean, minimum, and standard deviation of these values over the 10 runs were recorded. The results are listed in Table 2. (We provide a more detailed listing of the solution data along with solution plots in [28].) The computations were carried out in MATLAB on a 3 GHz computer with 3 GB of RAM.

On the small problems (1 through 5), our GA finds the optimal solution on nearly every run. The computation time was approximately 1 to 2 minutes per run. The exact algorithm required roughly 1.5 minutes for the small cases with 7 terminal nodes (problems 2, 4, and 5) and 20 minutes for those with 10 terminal nodes (problems 1 and 3). This is consistent with the Dreyfus-Wagner algorithm's complexity of about 3^N, where N is the number of terminal nodes. (Through numerical investigation, we estimate that our GA scales roughly like N^2.) Our GA outperforms the standard GA, which does not regularly find the optimal solution, by an average of about 4%. Run times for the two GAs were comparable.

For the medium-sized problems (6 through 10), our GA finds near-optimal solutions that are within around 1% of optimal on average. We present examples in Figures 8 through 11. These plots show that even when the GA solution is not optimal, it reproduces the most important structural features of the optimal solution in the way that it avoids high cost regions. The results show that our GA works well on problems with either structured or random input data (grids and nodes sets). Our GA was significantly faster than the exact algorithm on these medium-sized problems: 15 minutes versus 12 days. And it outperformed the standard GA by 14% on average, with similar run times. For problems 1 to 10, our GA (best run) was, on average, 0.19% above optimality, whereas the standard GA (best run) was, on average, 7.25% above optimality.

As mentioned above, problems 11, 12, and 13 were too large to be solved with the exact algorithm in a reasonable amount of time. But we do compare the two GAs on these problems and find that our GA (mean) remains superior to the standard GA (mean) by an average of around 6%. Here, as throughout the results, our GA exhibits a lower standard deviation over the 10 runs per problem than the standard GA, indicating it is also the more stable algorithm. One issue that becomes a more serious concern for our GA with increasing problem size is run time. The determination of the shortest path minimal spanning trees begins to slow the algorithm down compared to the standard GA. This effect becomes apparent in problem 13, which is the largest problem we solve. In that problem, the standard GA takes about one hour while our GA takes about 40 minutes for the GA to run, but with an additional 8 hours to set up the shortest paths the first time a problem is run with a particular grid cost structure. This amount of time could be significantly reduced, since at present, we compute the shortest paths between all pairs of possible node locations. In practice, these could be computed as needed. We are also studying other ways to reduce the run time of our algorithm, such as a divide and conquer approach.

5. CONCLUSION

In conclusion, we developed a genetic algorithm that finds optimal or near-optimal solutions for a variety of problem instances of the Euclidean Non-Uniform Steiner tree problem. The GA scales much better with problem size

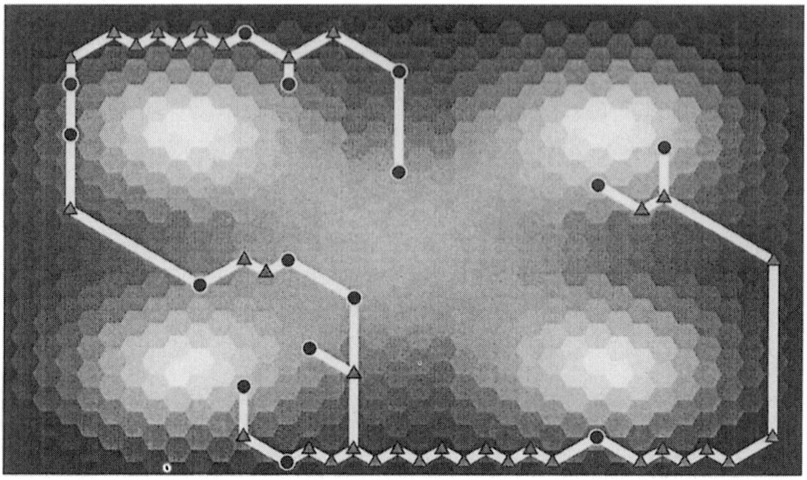

Figure 8. Optimal solution for Problem 6, Cost = 26.606. Triangles are Steiner nodes, black circles are terminal nodes. The lighter the cell, the more costly it is.

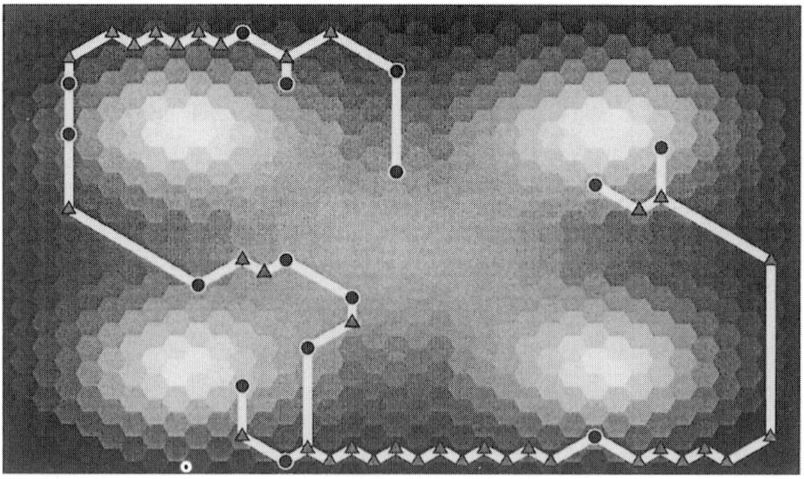

Figure 9. Sample solution found by our GA for Problem 6, Cost = 26.653. Triangles are Steiner nodes, black circles are terminal nodes. The lighter the cell, the more costly it is.

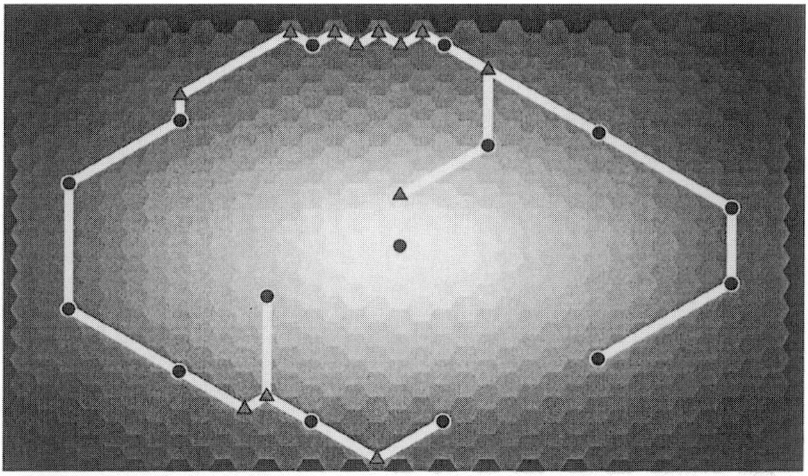

Figure 10. Optimal solution for Problem 9, Cost = 32.744. Triangles are Steiner nodes, black circles are terminal nodes. The lighter the cell, the more costly it is.

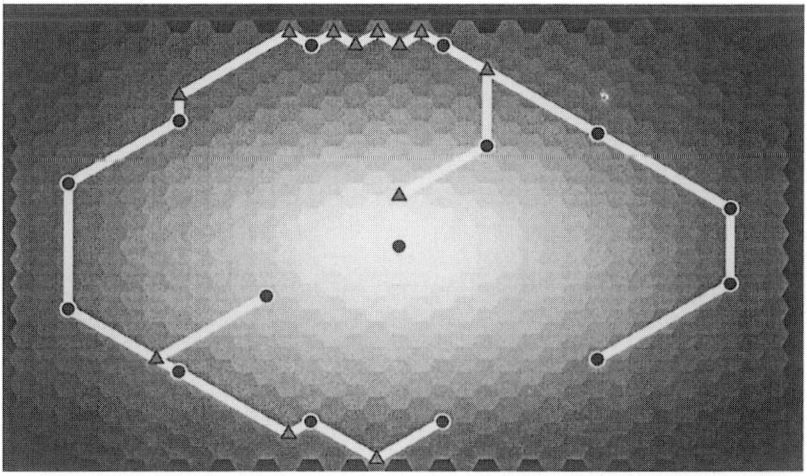

Figure 11. Sample solution found by our GA for Problem 9, Cost = 33.028. Triangles are Steiner nodes, black circles are terminal nodes. The lighter the cell, the more costly it is. Our GA was usually able to find the optimal solution for this problem.

than the exact algorithm to which we compare. By seeding our initial population with shortest-path minimal spanning trees, we were able to make considerable improvements over our earlier GA. In addition, our GA outperforms a different GA that uses more standard components.

Future extensions of this work could include improving the algorithm by using smarter, simpler operators, running it on node-weighted problems, and modifying it for rectilinear non-uniform problems. The algorithm could also be applied to problems utilizing geographic information systems (GIS) data to determine the grid cost structure and node locations. For example, the GIS data might factor steepness and vegetation into the grid cost determination [33, pages 142–147]. Or, as in the case of a recent study [34] on routing trucks bearing hazardous materials, the factors could include population counts, accident probability, risk of hijack, traffic conditions, emergency response, etc.

References

[1] P. Winter. The Steiner problem in networks: A survey. *Networks*, 17(2):129–167, 1987.

[2] J. M. Smith and J. S. Liebman. Steiner trees, Steiner circuits, and the interference problem in building design. *Engineering Design*, 4:15–36, 1979.

[3] C. Alpert, G. Gandham, J. Hu, J. Neves, S. Quay, and S. Sapatnekar. Steiner tree optimization for buffers, blockages and bays. *IEEE Transactions on Computer-Aided Design of Integrated Circuits and Systems*, 20(4):556–562, 2001.

[4] F. K. Hwang, D. S. Richards, and P. Winter. *The Steiner Tree Problem.* Number 53 in Annals of Discrete Mathematics. North-Holland, 1992.

[5] F. K. Hwang. A primer of the Euclidean Steiner tree problem. *Annals of Operations Research*, 33:73–84, 1991.

[6] S. L. Hakimi. Steiner's problem in graphs and its implications. *Networks*, 1:113–133, 1971.

[7] H. J. Prömel and A. Steger. *The Steiner Tree Problem: A Tour through Graphs, Algorithms, and Complexity.* Vieweg, Braunschweig/Wiesbaden, Germany, 2002.

[8] P. A. Thurber and G. Xue. Computing hexagonal Steiner trees using PCX. In *Proceedings of the International Conference on Electronics, Circuits and Systems*, pages 381–384, 1999.

[9] C. Coulston. Steiner minimal trees in a hexagonally partitioned space. *International Journal of Smart Engineering System Design*, 5:1–6, 2003.

[10] M. Zachariasen and P. Winter. *Workshop on Algorithm Engineering and Experimentation, Lecture Notes in Computer Science, Volume 1619*, Chapter: Obstacle-avoiding Euclidean Steiner trees in the plane: An exact algorithm, pages 282–295. Springer-Verlag GmbH, 1999.

[11] I. Frommer, B. Golden, and G. Pundoor. *The Next Wave in Computing, Optimization, and Decision Technologies*, Chapter: Heuristic methods for solving Euclidean non-uniform Steiner tree problems, pages 133–148. Springer, New York, 2005.

[12] Y. Kanemoto, R. Sugawara, and M. Ohmura. A genetic algorithm for the rectilinear Steiner tree in 3-D VLSI layout design. In *Proceedings of the 47th IEEE International Midwest Symposium on Circuits and Systems*, pages I–465 – I–468, 2004.

[13] C. Stanton and J. M. Smith. Steiner trees and 3-D macromolecular conformation. *INFORMS Journal on Computing*, 16(4):470–485, 2004.

[14] R. Karp. *Complexity of Computer Computations*, Chapter: Reducibility among combinatorial problems, pages 85–103. Plenum Press, New York, 1972.

[15] M.R. Garey, R.L. Graham, and D.S. Johnson. The complexity of computing Steiner minimal trees. *SIAM Journal on Applied Mathematics*, 32:835–859, 1977.

[16] C. Gröpl, S. Hougardy, T. Nierhoff, and H. J. Prömel. *Steiner Trees in Industry*, Chapter: Approximation algorithms for the Steiner tree problem in graphs. Kluwer Academic Publishers, Dordrecht, The Netherlands, 2001.

[17] J. Barreiros. An hierarchic genetic algorithm for computing (near) optimal Euclidean Steiner trees. In *Proceedings of the Genetic and Evolutionary Computation Conference (GECCO)*, pages 56–65, 2003.

[18] C. C. Ribeiro and M. C. de Souza. Tabu search for the Steiner problem in graphs. *Networks*, 36(2):138–146, 2000.

[19] L. J. Osborne and B. E. Gillett. A comparison of two simulated annealing algorithms applied to the directed Steiner problem on networks. *ORSA Journal on Computing*, 3:213–225, 1991.

[20] B. A. Julstrom. A scalable genetic algorithm for the rectilinear Steiner problem. In *Proceedings of the 2002 Congress on Evolutionary Computation*, pages 1169–1173, 2002.

[21] B. A. Julstrom. A hybrid evolutionary algorithm for the rectilinear Steiner problem. In *Proceedings of the Genetic and Evolutionary Computation Conference (GECCO)*, pages 49–55, 2003.

[22] C. Duin. *Steiner's problem in graphs*. PhD thesis, University of Amsterdam, Amsterdam, the Netherlands, 1993.

[23] G. Robins and A. Zelikovsky. Improved Steiner tree approximation in graphs. In *Symposium on Discrete Algorithms*, pages 770–779, 2000.

[24] T. Koch and A. Martin. Solving Steiner tree problems in graphs to optimality. *Networks*, 32:207–232, 1998.

[25] T. Polzin and S. Vahdati. Improved algorithms for the Steiner problem in networks. *Discrete Applied Mathematics*, 112(1-3):263–300, 2001.

[26] S. Dreyfus and R. Wagner. The Steiner tree problem in graphs. *Networks*, 1:195–207, 1972.

[27] J. Cheriyan and R. Ravi. *Lecture Notes on Approximation Algorithms for Network Problems*, chapter 7, Approximation algorithms for Steiner trees, pages 92–95. `http://www.gsia.cmu.edu/afs/andrew/gsia/ravi/WWW/new-lecnotes.html`.

[28] I. Frommer. *Modeling and Optimization of Transmission Networks*. PhD thesis, University of Maryland, College Park, MD, USA, 2005.

[29] M. Mitchell. *An Introduction to Genetic Algorithms*. The MIT Press, Cambridge, MA, 1998.

[30] R. Floyd. Algorithm 97 (SHORTEST PATH). *Communications of the Association for Computing Machinery*, 5:345, 1962.

[31] R. Prim. Shortest connection networks and some generalizations. *Bell Systems Technical Journal*, 36:1389–1401, 1957.

[32] S. H. Jung. Queen-bee evolution in genetic algorithms. *IEEE Electronic Letters*, 39:575–576, 2003.

[33] A. Mitchell. *The ESRI Guide to GIS, Volume 1: Geographic Patterns and Relationships.* ESRI Press, Redlands, California, 1999.

[34] B. Huang, R. L. Cheu, and Y. S. Liew. GIS and genetic algorithms for HAZMAT route planning with security considerations. *International Journal of Geographical Information Science*, 18(8):769–787, 2004.

CARDINALITY AND THE SIMPLEX TABLEAU FOR THE SET PARTITIONING PROBLEM

Anito Joseph and Edward K. Baker
Department of Management Science, University of Miami, Coral Gables, Florida 33124, ajoseph@miami.edu and ebaker@miami.edu

Abstract: In this work, we show how cardinality-related information for the set partitioning problem is represented within the simplex tableau and how a fractional solution can be interpreted in terms of unresolved solution cardinality. We include a cardinality row within the linear programming relaxation of the set partitioning problem to demonstrate the associated cardinality-related information present in the tableau. Working with a basic feasible solution, the cardinality row is shown to provide valuable information for branching along the cardinality dimension of the solution space of the problem. It is shown that cardinality information may be derived from the simplex tableau for any subset of structural variables in the problem. An illustrative example and computational results for problems from the literature are presented.

Key words: cardinality, simplex tableau, set partitioning.

1. INTRODUCTION

The set partitioning problem (SPP) considers a set of m elements, {M}, that have been grouped into n distinct subsets, P_j, j = 1, 2,..., n. Typically a cost, c_j, is associated with each subset P_j. For example, if the original m elements are tasks to be performed, then each P_j may be considered a possible subset of tasks that may performed by a single actor and c_j would be the cost of having that actor perform those tasks. The set partitioning problem is then to find the minimal cost selection of mutually exclusive subsets P_j such that the union of the selected subsets is equal to {M}.

The set partitioning problem is modeled as the following binary linear program:

Minimize $\Sigma\ c_j\ x_j$,

Subject to: $\Sigma\ a_{ij}\ x_j\ = 1, i = 1, 2, \ldots, m$,

x_j ? $\{0,1\}, j = 1, 2, \ldots, n.$

In the formulation, a set of binary decision variables, $\{x_j\}$, is associated with the n subsets, P_j, used in the model. In the solution process $x_j = 1$ if subset P_j is used in the solution; $x_j = 0$ otherwise. The cost of each subset P_j is denoted c_j and, in the constraint set, $a_{ij} = 1$ if the i-th element of the set M is contained in the j-th subset; $a_{ij} = 0$ otherwise.

The cardinality of a set is equal to the number of elements contained in that set. There are numerous combinatorial optimization problems that are concerned with determining minimal or maximal cardinality solutions. These include the problems of determining covers, colorings, and cliques, as well as, problems of cardinality constrained tours and schedules. The classic book of Garey and Johnson (1979) lists several such examples and shows many of these problems to be NP-hard. Recent work on cardinality constrained optimization problems has included portfolio selection (Bienstock (1996) and Chang et al. (2000)), the traveling salesman problem (Cao and Glover (1997) and Patterson and Roland (2003)), and capacitated minimal spanning trees (Gouveia and Martins (2005)).

Solution cardinality is also an active stream of research in constraint programming. Fages and Lal (2006), for example, have applied a constraint programming approach to cardinality constrained cutset problems. Additional constraint programming methods have been proposed for various cardinality constrained routing and scheduling problems (Azevedo (2003), Regin (2001), and Smith et al. (2001)), as well as to general integer programming problems (Zabatta (2001)).

The focus of this paper is the extraction of cardinality-related information from the simplex tableau of the linear programming relaxation of the set partitioning problem. The consideration of cardinality within the simplex tableau has been of interest to researchers for some time (see for example Rubin (1974)). The recent work of Joseph and Baker (2006), however, has focused on cardinality probing and the explicit use of a cardinality constraint within the set partitioning framework.

In the remainder of this paper, we develop the theoretical relationship between the simplex tableau and the cardinality constraint. In Section 2, we demonstrate how cardinality-related information can be obtained by augmenting the set partitioning problem with a cardinality row. The cardinality row of the simplex tableau is shown to provide valuable information for branching along the cardinality dimension of the solution space of the problem. It is also shown that cardinality-related information may be derived from the LP solution for any subset of structural variables in the problem. In Section 3, an illustrative example and computational results for set partitioning problems found in the literature are presented. The paper concludes in Section 4 with the conclusions of this research and suggestions for future work.

2. THE SIMPLEX TABLEAU AND CARDINALITY

Consider the LP relaxation of the set partitioning problem P0: $\{\text{Min } cX|AX = e, X = 0\}$ where e is column vector of ones. Without loss of generality, the matrix, A, and the vector of variables, X, may be partitioned into their basic and nonbasic variable components, such that $A = (B|N)$ and $X = (X_B|\ X_N)$. The problem P0 may then be written as $\{\text{Min } cX|BX_B = e - NX_N, X=0\}$. Solving P0, we find at LP optimality, $Z^* = c_BX_B + c_NX_N$, where $X_B = B^{-1}\mathbf{e} - B^{-1}NX_N$. Let $B^{-1}\mathbf{e} = \underline{b}$. Note that the operation $B^{-1}\mathbf{e}$ sums the elements of each row of B^{-1} to arrive at the elements of $\underline{b}$.

To obtain the sum of the basic variables, i.e. the sum of the elements of $\underline{b}$, the operation $\mathbf{e}'\underline{b} = \mathbf{e}'(B^{-1}\mathbf{e}) = (\mathbf{e}'B^{-1})\mathbf{e}$ is performed. Therefore, with $(\mathbf{e}'B^{-1})$, we first sum the columns of B^{-1} to form a row vector, then the operation $(\mathbf{e}'B^{-1})\mathbf{e}$ sums the elements of the row vector to find the sum of the basic variable values. Alternatively, the operation $(B^{-1}\mathbf{e})$ sums the rows of B^{-1} to give a column vector, then the operation $\mathbf{e}'(B^{-1}\mathbf{e})$ finds the sum of the basic variable values by summing the rows of the column vector $B^{-1}\mathbf{e}$. Note that if the elements of $\underline{b}$ are integer, then the sum $\mathbf{e}'\underline{b}$ is the actual cardinality of the solution, i.e. the number of structural variables in the solution. Since the integrality of the LP solution is not guaranteed, we refer to the sum $\mathbf{e}'\underline{b}$ as a pseudo-cardinality, an approximation of structural variable cardinality.

To assess the effect of changes in the nonbasic variables on the basic variable values, we note that from the final LP tableau each basic variable may be expressed as $X_B = \underline{b} - (B^{-1}N)X_N$. For any individual nonbasic variable x_j, and its associated column in the matrix A, A_j, let $B^{-1}A_j = \bar{A}_j = \{\bar{a}_{ij}\}$. Then for the i-th element of X_B, $x_{Bi} = \underline{b}_i - \bar{a}_{ij}\theta_j$ where θ_j is the value assumed by the nonbasic variable x_j. Disregarding feasibility concerns, if a nonbasic variable x_j is changed from a value of zero to a value of θ_j, the values $\{\bar{a}_{ij}\}$ provide the corresponding changes in the X_B values. Hence, the cardinality approximation, $\mathbf{e}'X_B$, is changed by the amount

$$-(?_i \bar{a}_{ij}\theta_j) + \theta_j,$$

where θ_j is the value assumed by the nonbasic variable x_j. When $\theta_j = 1$, this change is equal to

$$-(?_i \bar{a}_{ij}) + 1.$$

If we interpret the SPP as a problem of unresolved cardinality, we can reframe our problem as one of correcting or refining the pseudo-cardinality to approach a solution with a particular cardinality of minimum cost.

2.1 Augmentation for Cardinality

The cardinality, or pseudo-cardinality, of the current LP solution may be monitored explicitly by adding the equation

$$?_j x_j - y = 0$$

as the m+1st row in problem P0. Using the notation of the previous section, we define this augmented problem as P1: $\{\text{Min } c\mathbf{X} | \mathbf{AX} = e,\ ? \ x_j - y = 0;\ \mathbf{X} = 0,\ y = 0\}$, where y represents the pseudo-cardinality. Solving problem P1 will give the same basic solution as problem P0 with the addition of the variable y. We will show that the row of the LP tableau for problem P1 associated with the basic variable y will indicate the effect of each nonbasic variable on the pseudo-cardinality.

Let T and T^{-1} be the basis and basis inverse matrices, respectively, for problem P1. Then T is simply the matrix B extended by one additional row and column. Row m+1 of T contains the value 1 for the first m columns and the value - 1 for the m+1st column. Column m+1 of T contains the value zero for the first m rows and the value - 1 in the m+1st row.

The basis inverse matrix, T^{-1}, can be easily obtained from the augmented matrix T and knowledge of the original basis inverse B^{-1}. Since $T^{-1}T = \mathbf{I}$, the first m rows and m columns of T^{-1} must equal B^{-1}. Additionally, we know that the m+1st column of **I** must consist of zeros as the first m elements and the value one as the m+1st element. Thus, for elements 1 to m of the m+1st column of **I**, $\sum_{j<(m+1)} T^{-1}_{i,j} * (0) + T^{-1}_{i,(m+1)} *(-1) = 0$. Following the algebra, this reduces to: $0 + T^{-1}_{i,(m+1)} *(-1) = 0$. Hence, $T^{-1}_{i,(m+1)} = 0$ for rows 1 to m.

In a similar manner, the element in row m+1 of column m+1 of T^{-1} is equal to: $?_{j<(m+1)} T^{-1}_{(m+1),j} *(0) + T^{-1}_{(m+1),(m+1)} *(-1) = 1$. Simplifying, we find this reduces to $0 + T^{-1}_{(m+1),(m+1)} *(-1) = 1$. Hence, $T^{-1}_{(m+1),(m+1)} = -1$.

Consider the matrix multiplication to determine the elements of row m+1 of **I**, involving row (m+1) of T and column j of T^{-1}. Recall that the first m elements of the m+1st row of T all equal 1 and the m+1st element of the row is -1. This operation effectively sums the first m elements of column j of T^{-1}, ? $_iB^{-1}_{i,j}$, (i.e., the elements of column j in B^{-1}), and then subtracts $T^{-1}_{(m+1),j}$, the element in row m+1 of column j of T^{-1}. Now ? $_iB^{-1}_{i,j}$ - $T^{-1}_{(m+1),j}$ = 0 for the first m elements of row m+1 of **I**, and therefore, for the associated columns, ? $_iB^{-1}_{i,j} = T^{-1}_{(m+1),j}$, j = m. In the matrix notation of the previous section, the row vector of m elements $[T^{-1}_{(m+1),j}$, j =1, ..., m]'= **e**' B^{-1}. The elements $T^{-1}_{(m+1),j}$ show the overall contribution of each of the associated columns of B^{-1} to **e**'b, the sum of the elements of b, (**e**' B^{-1}**e** =**e**'b = (**e**' B^{-1})**e**).

Note that, in the augmented problem, the matrix of nonbasic columns, N, now has m+1 rows where the elements of the m+1st row are all ones. Recalling that N = $\{N_{ik}\}$ consists of a matrix of 0's and 1's, so that the operation $T^{-1}N$ = A, means that we are conducting a series of summations that involve either adding an element of T^{-1} (N_{ik} = 1) or not adding it (N_{ik} = 0) to the current sum. For any row i, i < (m+1), the element in the kth column of A, a_{ik}, shows how the kth nonbasic variable would affect the value of basic variable in row i if this nonbasic variable were to enter the basis at value 1 (disregarding feasibility). Following the notation of the previous section, $[T^{-1}_{(m+1),j}$, j =1, ...m]' = **e**' B^{-1}, and finding the product of this row vector and the first m rows of A_k gives **e**' $B^{-1}A_k$ = ? a_{ik}. Except for notation, this is exactly the first term in the expression for of the cardinality effect for nonbasic variable x_j identified earlier in Section 2. To obtain the actual effect of - (? a_{ik}.- 1), recall that $T^{-1}_{(m+1),(m+1)}$ = - 1 and the coefficient value in the m+1st row of matrix N, or the m+1st row of column A_k is equal to 1, so that $[T^{-1}_{(m+1),j}$, j =1, ...m | - 1]'A_k = **e**' $B^{-1}A_k$ + (-1)*(1) = ? a_{ik}.- 1. Since X_B= b - $(B^{-1}N)X_N$, where the cardinality effect is given by - $(B^{-1}N)X_N$, the cardinality effect for the augmented problem is simply - $(T^{-1}N)X_N$, then the cardinality effect for the m+1st row becomes - (? a_{ik}- 1) .

2.2 An Illustrative Example

Consider the problem P0 with matrix A and objective function vector c.

c = (72 62 75 50 58 52 59 55)

A = 0 1 0 0 1 0 0 1

0 1 0 0 0 1 0 1

0 0 0 1 0 1 1 0

1 1 1 1 0 0 0 0

1 0 0 0 1 0 1 0

0 0 1 0 0 1 0 0

The LP solution, X0 = (.25, .25, .25, .25, .75, .75, 0, 0). Hence, the pseudo-cardinality of this solution is 2.5. Note that this value is not the objective function value for this weighted set partitioning problem, but rather the simple sum of the basic variable values.

The inverse matrix, B^{-1} equals

-.75	.5	-.25	.25	.75	-.25
.25	.5	-.25	.25	-.25	-.25
.25	-.5	-.25	.25	-.25	.75
.25	-.5	.75	.25	-.25	-.25
.75	-.5	.25	-.25	.25	.25
-.25	.5	.25	-.25	.25	.25

Hence $B^{-1}N$ would equal

.5	-.25
-.5	.75
-.5	-.25
.5	-.25
.5	.25
.5	.25

If a redundant row ? x_j - y = 0 is included to measure the pseudo-cardinality, y, and then the augmented matrix A becomes:

0	1	0	0	1	0	0	1	0
0	1	0	0	0	1	0	1	0
0	0	0	1	0	1	1	0	0
1	1	1	1	0	0	0	0	0
1	0	0	0	1	0	1	0	0
0	0	1	0	0	1	0	0	0
1	1	1	1	1	1	1	1	-1

This corresponds to the T matrix mentioned in the previous section.

The inverse basis matrix, T^{-1} is given by:

-.75	.5	-.25	.25	.75	-.25	0
.25	.5	-.25	.25	-.25	-.25	0
.25	-.5	-.25	.25	-.25	.75	0
.25	-.5	.75	.25	-.25	-.25	0
.75	-.5	.25	-.25	.25	.25	0
-.25	.5	.25	-.25	.25	.25	0
.50	0	50	.50	.50	.50	-1

Hence $T^{-1}N$ would now be equal to:

.5	-.25
-.5	.75
-.5	-.25
.5	-.25
.5	.25
.5	.25
0	-.5

Thus for x_7 (first column of $B^{-1}N$ and $T^{-1}N$), the overall effect on the pseudo-cardinality is given by – (? $a_{ij}\theta_7$) + θ_7 = - (1) (θ_7) + θ_7 = 0. For x_8 (second column of $B^{1}N$ and $T^{1}N$), it means an overall effect of - (.5) (θ_8) + θ_8 = .5θ_8. When $\theta_7 = 1$, the overall effect is 0, that is no effect on the current value of pseudo-cardinality; when $\theta_8 = 1$, the overall effect = .5, an increase of .5 in pseudo-cardinality.

3. FINDING AN INTEGER SOLUTION IN THE CARDINALITY DIMENSION

Given a cardinality row, ? x- y = 0, we can determine the effect of all nonbasic variables on y, the pseudo-cardinality of the basic variables. The "cardinality" effect is additive so that in searching for a feasible integer solution the question is to identify the nonbasic column, or columns, that correct the pseudo-cardinality to a true cardinality and result in a feasible integer solution.

Solution cardinality can be used as a factor in the search for an integer solution by searching in either direction from the LP solution pseudo-cardinality. Thus in the direction of increasing cardinality, at least one of the nonbasic variables with an increasing effect must enter the basis, and vice versa for a search in the direction of decreasing cardinality. While we are still in the early stages of exploring solution strategies for using cardinality effect information, we know that traditional techniques for identifying nonbasic variables to enter the basis, e.g. cutting planes and enumeration along the lines used in Joseph and Baker (2006) can be employed. The research question remaining is to determine how to obtain maximum benefit from using cardinality information.

3.1 Variable Subsets

While the methods in Section 2 were developed for the entire set of structural variables, the analysis also extends to any subset of structural variables, for example, a clique. Thus if a clique is identified, the LP simplex tableau provides the effect of the nonbasic variables on the clique pseudo-cardinality. In the past, researchers have searched for violated cliques to exploit in their search for a solution, but here the cardinality effect provides an opportunity to use any clique (violated or not) in the search for an integer solution. Because the relevant information for any variable subset can be readily constructed from the LP simplex tableau, we may employ solution strategies where variable subsets are identified prior to solution or variable subsets with unresolved cardinalities are identified post LP solution.

3.2 Computational Problems from the Literature

We briefly discuss examples using the solution cardinality considerations to search for an integer solution. The problems used for the computational results were taken from the set partitioning problems used in the paper of Hoffman and Padberg (1993).

The first problem considered has 197 columns and 17 rows. The LP relaxation has the solution: $Z0 = 10972.5$, with $y = 4.5$. The pseudo-cardinality y has a fractional value and we use a Gomory (1963) cut on the pseudo-cardinality to improve the LP solution. The new solution is integer with $Z = 11307$, and $y = 5$. This is an optimal solution.

An alternative approach is to use a branch search. Since y is fractional, we can branch on integer values in either direction from y, that is, search at $y = 5$ and $y = 4$. When $y = 5$, the solution is integer with $Z = 11307$and $y = 5$. When $y = 4$, the solution is non-integer with $Z = 13444.5$, and $y = 4$.

For this small problem, an optimal solution was quickly found by searching solely on the problem pseudo-cardinality, y. This is not the typical result, however, and additional variable subsets may have to be identified to help in the search.

The second problem considered has 685 columns and 22 rows. Solving the linear programming relaxation problem, we find that Z0 = 16626. The solution is fractional, yet the pseudo-cardinality, y = 5.

Since the pseudo-cardinality is integer, we cannot use the y value to obtain a Gomory cut. In this case, we need to identify one or more variable subsets upon which to branch. This can be done by splitting the basic structural variables into two or more subsets that have fractional valued pseudo-cardinalities. Gomory cuts or branching can be applied to search for an integer solution, as was done for the previous example. For this example, however, we will identify other variable subsets for use in our search.

For this problem we will attempt to search for an integer solution at the cardinality of 5. To identify variable subsets, we revert to early approaches for solving the SPP. In early implicit enumeration approaches to the set partitioning problem, researchers (see for example Garfinkel and Nemhauser (1969), Marsten (1974), Pierce (1968) and Pierce and Lasky (1973)) exploited the problem structure by partitioning the matrix into a number of blocks. In this approach, rows and columns of the matrix A are rearranged and partitioned into blocks, such that column j is placed in block k if $k = \min\{i|a_{ij}=1\}$. Doing this for problem 685x22, we identify 18 blocks described in the table below. Each block identifies a unique clique of structural variables.

Indexing these subsets as S_k, k =1, ..., 18, we can augment the problem, implicitly or explicitly, with 18 rows in a similar format to the general cardinality row ? x- y = 0; thus ? $_jx_j$ - $S_k = 0$, j ? x^K, where x^K represents the subset of structural columns belonging to subset S_k. Thus when S_k is equal to 0, it means that none of the structural variables in that clique is active in the solution. When S_k is positive then it means that the clique is active in the solution, but activity is not necessarily integer-valued.

The LP solution gives $S_1= S_2= S_3=1$, $S_6 = S_{10} =.75$, and $S_7 = S_8 = .25$. We will use the fractional subsets S_6, S_7, S_8, and S_{10}, to search for an integer solution. Since the cardinality of these subsets cannot exceed 1, then the cardinality effect of the nonbasic variables on these clique subsets is restricted to an absolute value between 0 and 1 or more strictly between S_k and (S_k - 1). We will use the relation between the basic subset S_k and the nonbasic structural variables {$S_k = \underline{b}_k$ - (? $_j\bar{a}_{kj}\theta_j$) = 0, k = 1,..., 18} to obtain cardinality effects: $\underline{b}_k$ = (? $_j\bar{a}_{kj}\theta_j$) for each fractional subset. These effects coupled with the equirement that at least one of the nonbasic variables identified must enter the basis will be used for the search. In order to coax

Problem 685x22 – Block Information

Block Number	Row	Beginning Column
1	1	1
2	2	9
3	3	45
4	4	79
5	5	107
6	6	151
7	7	246
8	8	331
9	9	421
10	10	472
11	11	562
12	12	587
13	13	601
14	14	615
15	16	643
16	17	667
17	21	684
18	22	685

the search to remain at the true cardinality of 5, we select only nonbasic columns with zero effect on the pseudo-cardinality. Though not necessary, we also limit the reduced cost of the nonbasic columns selected to not exceed 600.

Using these four cardinality effects and the requirement that at least one of the nonbasic variables identified should enter the basis, produced an integer solution of cardinality 5 and an objective function value of 16812. This is the optimal solution for this problem which can be confirmed by including all nonbasic variables in the four cardinality effect relations.

These two examples above illustrate the many possibilities for using cardinality in the search for an integer solution. We intend to develop and test several algorithms that exploit the cardinality information to narrow the duality gap and bound or solve the problem.

4. CONCLUSIONS AND DIRECTIONS FOR FUTURE RESEARCH

In this paper, we have described how solution cardinality information may be generated from within the LP tableau and we have shown how this information can be obtained for any subset of structural variables. The subset of structural variables does not have to be explicitly identified prior to finding the LP solution. Using cardinality allows information about any subset with unresolved cardinality (i.e. subsets with fractional-valued variables) to be used in the search for an integer solution rather than having to search for subsets that violate some particular relationship, e.g., cliques. We suggest various techniques for identifying nonbasic variables to enter the basis in the search for an integer solution. Future research will examine areas such as subset identification, cutting planes, and enumeration approaches to determine how best to employ underlying cardinality information in identifying feasible and optimal integer solutions for SPP. The results of our exploration will be used to develop algorithms that exploit cardinality to reduce the duality gap and bound or solve SPP. In addition future research will investigate extensions of the cardinality approach to the set packing and set covering problems.

References

Azevedo, F., 2003. *Constraint Solving over Multi-valued Logics: Application to Digital Circuits*. IOS Press. Berlin.

Bienstock, D., 1996. Computational study of a family of mixed-integer quadratic programming problems. *Mathematical Programming* 74, 121-140.

Cao, B., and Glover, F., 1997. Tabu search and ejection chains – application to a node weighted version of the cardinality-constrained TSP. *Management Science* 43 (7), 908-913.

Chang, T-J., Meade, N., Beasley, J.E., and Sharaiha, Y.M., 2000. Heuristics for cardinality constrained portfolio optimization. *Computers and Operations Research* 27 (13), 1271-1302.

Fages, F., and Lal, A., 2006. A constraint programming approach to cutest problems. *Computers and Operations Research* 33 (10), 2852-2865.

Garey, M.R., and Johnson, D.S., 1979. *Computers and Intractability: A Guide to the Theory of NP-completeness*. W.H. Freeman and Company, San Francisco.

Garfinkel, R.S., and Nemhauser, G.L., 1969. The set-partitioning problem: set covering with equality constraints. *Operations Research* 17, 848-856.

Gomory, R., 1958. Outline of an algorithm for integer solutions to linear programs. Bulletin of the American Mathematical Society 64,275-278.

Gouveia, L., and Martins, P., 2005. The capacitated minimum spanning tree problem; revisiting hop-indexed formulations. *Computers and Operations Research* 32 (9), 2435-2452.

Hoffman, K., and Padberg M., 1993. Solving airline crew scheduling problems by branch-and-cut. *Management Science*, 39, 657-682.

Joseph, A., 2002. A concurrent processing framework for the set partitioning problem, *Computers and Operations Research*, 29, 1375-1391.

Joseph, A., and Baker, E.K., 2006. Parametric cardinality probing in set partitioning problems. Forthcoming in Perspectives in Operations Research: Papers in Honor of Saul Gass's 80th Birthday. Editors: Frank Alt, Michael Fu, and Bruce Golden.

Marsten, R.E., 1974. An algorithm for large set partitioning problems. *Management Science* 20, 774-787.

Patterson, R., and Rolland, E., 2003. The cardinality constrained covering traveling salesman problem. Computers and Operations Research 30 (1), 97-116.

Pierce,. J. F., 1968. Application of combinatorial to a class of all zero-one integer programming problems. *Management Science* 15, 191-209.

Pierce, J.F., and Lasky, J.S., 1973. Improved combinatorial programming algorithms for a class of all zero-one integer programming problems. *Management Science* 18, 528-543.

Regin, J., 2001. Minimization of the number of breaks in sports scheduling problems using constraint programming. In *Constraint Programming and Large Scale Optimization.* DIMACS 57, 115-130.

Rubin, D., 1974. Vertex generation and cardinality constrained linear programs. *Operations Research* 23 (3), 555-565.

Smith, B.N., Layfield, C.J., and Wren, A., 2001. A constraint pre-processor for a bus driver scheduling system. In *Constraint Programming and Large Scale Optimization.* DIMACS 57, 131-148.

Zabatta, F., 2001. Multithreaded constraint programming: a hybrid approach. In *Constraint Programming and Large Scale Optimization.* DIMACS 57, 41-64.

AN EFFICIENT ENUMERATION ALGORITHM FOR THE TWO-SAMPLE RANDOMIZATION DISTRIBUTION

Marie A. Coffin,[1] James P. Jarvis,[2] and Douglas R. Shier[2]

[1] *Biostatistics Group, Monsanto, Inc., Research Triangle Park, NC 27709*

[2] *Department of Mathematical Sciences, Clemson University, Clemson, SC 29634*

Abstract In many experimental situations, subjects are randomly allocated to treatment and control groups. Measurements are then made on the two groups to ascertain if there is in fact a statistically significant treatment effect. Exact calculation of the associated randomization distribution theoretically involves looking at all possible partitions of the original measurements into two appropriately-sized groups. Computing every possible partition is computationally wasteful, so our objective is to systematically enumerate partitions starting from the tail of the randomization distribution. A new enumeration scheme that only examines potentially worthwhile partitions is described, based on an underlying partial order. Numerical results show that the proposed method runs quickly compared to complete enumeration, and its effectiveness can be enhanced by use of certain pruning rules.

Keywords: Majorization, partition, permutation test, p-value, randomization test, two-sample test.

1. Introduction

Consider an experiment in which measurements are made on the members of two groups. On the basis of these measurements, we wish to ascertain whether or not the two groups arise from populations having the same mean. If the members of the two groups were chosen at random from their respective populations (and if the sample sizes are sufficiently large), one may appeal to the Central Limit Theorem to perform some version of a two-sample t-test on the null hypothesis $H_0 : \mu_1 = \mu_2$.

However, in many experimental situations, the members of the two groups are not chosen at random from given (normal) populations [7]. For example, in a designed experiment, the experimental units may be chosen for their convenience and/or availability, although some attempt to get a representative sample is often made. In such a case, the experimental units are assigned to one or the other of the treatment groups at random. The correct null distribution is actually the *randomization distribution*, which arises as follows.

Suppose that the treatment has, in fact, no effect on the response. Let $X_{11}, X_{12}, \ldots, X_{1n_1}$ and $X_{21}, X_{22}, \ldots, X_{2n_2}$ be the observations in the first and second samples, respectively. Under the null hypothesis, the observed difference in means $\overline{X}_1 - \overline{X}_2$ is induced only by the randomization — i.e., by the specific partitioning of experimental units into the two groups. Since every partition is equally likely, we can calculate exactly the probability of obtaining a difference $\overline{X}_1 - \overline{X}_2$ at least as large as that observed; thus we have the observed significance level (p-value) of our hypothesis test. Comprehensive coverage of this approach, in which minimal assumptions are made on the nature of the underlying populations, can be found in [3, 5, 8].

The exact randomization distribution depends on calculating $\overline{X}_1 - \overline{X}_2$ for every possible partition of the data. With n_1 observations in the first experimental group and n_2 observations in the second experimental group, there are $\binom{n_1+n_2}{n_1}$ such partitions. For example, if each group has 10 observations, there are 184,756 partitions to be calculated. If each group has 30 observations, the number of partitions exceeds 10^{17}. It has been shown [4, 12] that this randomization distribution, properly scaled, converges to the t-distribution. Because of this, calculating the huge numbers of partitions necessary for an exact p-value based on the randomization distribution with large samples is only necessary if the data indicate that the underlying distribution is severely nonnormal. However, even for moderate-sized samples and a high-speed computer, the calculations required to find all such partitions can take a significant amount of time. Monte Carlo methods can be used to sample from the randomization distribution and provide an approximate p-value. As an alternative, we show how the *exact* p-value (or critical value) can be calculated very quickly, making it feasible to use randomization tests for sample sizes in which the standard t-test assumptions may be suspect.

In some sense, generating every possible partition is computationally wasteful; to compute the p-value of the hypothesis test, the only partitions needed are those for which $\overline{X}_1 - \overline{X}_2$ is more extreme than the observed value. In this paper, we systematically generate partitions starting from the tail of the randomization distribution, terminating when the observed value has been passed, thereby reducing the required computational effort. In this way, we make mathematically precise the process originally suggested by Fisher [4] in the context of paired-sample tests. Diaconis and Holmes [2] use Gray codes as an alternative method to calculate p-values for exact paired-sample tests and mention that their approach can be extended to two-sample randomization tests. Pagano and Tritchler [11] use a fast Fourier transform method for two-sample problems, although the complexity of their algorithm is quite sensitive to the precision of the data. In addition, Mehta et al. [10] describe a dynamic programming approach on a layered network, in which each path corresponds to a partition. The efficacy of their approach depends on the number of distinct values of the test statistic.

By contrast, our approach exploits an underlying partial order, defined using the concept of majorization. Section 2 contains the mathematical framework underlying the efficient generation of partitions and introduces our state generation algorithm. Section 3 discusses various data structures and implementation details for efficiently carrying out this algorithm. Numerical results are also presented to indicate the computational effectiveness of the proposed procedure.

Note that by assumption the sample sizes and observed values are fixed: under the null hypothesis, the only variability arises from how the observed values are randomly assigned to the two groups. Thus the randomization test may be performed on $\overline{X}_1 - \overline{X}_2$, or on $\sum_{j=1}^{n_1} X_{1j} - \sum_{j=1}^{n_2} X_{2j}$, or simply on $\sum_{j=1}^{n_1} X_{1j}$. These are equivalent since the sum of all observations is fixed. For simplicity in presentation, we choose the last of these quantities as our test statistic.

2. State Generation Approach

Suppose that observations $X_1, X_2, \ldots, X_n$ are made on $n = n_1 + n_2$ units, with $k = n_1$ of the units belonging to Group 1 and the remaining $n - k = n_2$ units belonging to Group 2. Under the null hypothesis that all observations derive from the same population, every partition of units into k of Group 1 and $n - k$ of Group 2 is equally likely. If the k units of Group 1 are indexed by $I = \{i_1, i_2, \ldots, i_k\}$ then the sum of observations (or *value*) associated with this index set is given by

$$v(I) = \sum_{i \in I} X_i.$$

The objective then is to calculate tail probabilities for the distribution of $v(I)$, namely, $\Pr(v(I) \geq v_o)$. Rather than generating all $\binom{n}{k}$ partitions of the n units and calculating the distribution directly, the indirect approach taken here begins with the partition having the most extreme value for $v(I)$ and then successively generates less extreme values until an appropriate cutoff value v_o for the test statistic has been reached. In this way, it is only necessary to consider values likely to be in the extreme tail of the distribution. Many partitions will never be encountered and this should lead to tangible computational savings. The present section discusses the mathematical framework for such an approach.

To begin, assume that the n observations X_i have been placed in nonincreasing order

$$X_1 \geq X_2 \geq \cdots \geq X_n. \tag{1}$$

Our approach will generate k-subsets I of $\{1, 2, \ldots, n\}$ — that is, subsets I with $|I| = k$ — in nonincreasing order of their value $v(I)$. By virtue of (1), it is clear that the largest value of $v(I)$ occurs for $I = \{1, 2, \ldots, k\}$. Moreover, the inequality (1) implies a definite relationship between the values for certain k-subsets I and J. For example, if $I = \{1, 2, 4\}$ and $J = \{2, 3, 5\}$ then

$$v(I) = X_1 + X_2 + X_4 \geq X_3 + X_2 + X_4 \geq X_3 + X_2 + X_5 = v(J).$$

If we can succinctly capture this (partial) ordering information among k-subsets, then we can implicitly select a k-subset I with the largest current value $v(I)$ without having to consult all possible k-subsets.

With this end in mind, define the *state vector* associated with the k-subset $I = \{i_1, i_2, \ldots, i_k\}$ to be the binary n-vector having a 1 in positions $i_1, i_2, \ldots, i_k$ and 0's elsewhere. That is, the k units of Group 1 appear in positions having a 1 and the $n - k$ units of Group 2 appear in positions having a 0. Notice that if $\boldsymbol{s} = (s_1, s_2, \ldots, s_n)$ is the state vector corresponding to I, then

$$\sum_{i=1}^{n} s_i = k \quad \text{and} \quad v(\boldsymbol{s}) \equiv v(I) = \sum_{i=1}^{n} s_i X_i.$$

We can define a partial ordering $\succeq$ on states by

$$\boldsymbol{s} \succeq \boldsymbol{t} \quad \Leftrightarrow \quad \sum_{i=1}^{r} s_i \geq \sum_{i=1}^{r} t_i \ \text{ for all } r = 1, 2, \ldots, n. \tag{2}$$

This order relation is essentially the *majorization order* [9] on n-vectors having nonnegative components summing to a common value (here k).

LEMMA 1 $\boldsymbol{s} \succeq \boldsymbol{t} \Leftrightarrow v(\boldsymbol{s}) \geq v(\boldsymbol{t})$ *for all* X_i *satisfying (1).*

Proof. Assume $\boldsymbol{s} \succeq \boldsymbol{t}$ and suppose that $I = \{i_1, i_2, \ldots, i_k\}$, $J = \{j_1, j_2, \ldots, j_k\}$ are the index sets associated with states $\boldsymbol{s}, \boldsymbol{t}$ respectively. Using definition (2) with $r = j_1$ it follows that $i_1 \leq j_1$. In a similar way it follows in succession that $i_2 \leq j_2, \ldots, i_k \leq j_k$. Thus by the ordering assumption (1)

$$v(\boldsymbol{s}) = \sum_{r=1}^{k} X_{i_r} \geq \sum_{r=1}^{k} X_{j_r} = v(\boldsymbol{t}).$$

The reverse implication is easily established by considering the particular selection $X_1 = \cdots = X_r = 1$ and $X_{r+1} = \cdots = X_n = 0$. Then $v(\boldsymbol{s}) \geq v(\boldsymbol{t})$ means

$$\sum_{i=1}^{r} s_i = \sum_{i=1}^{n} s_i X_i \geq \sum_{i=1}^{n} t_i X_i = \sum_{i=1}^{r} t_i$$

and so $s \succeq t$. □

To illustrate, if $n = 5$ and $k = 3$ then $(1, 1, 0, 1, 0) \succeq (0, 1, 1, 0, 1)$ since the partial sums $1, 2, 2, 3, 3$ for s dominate the partial sums $0, 1, 2, 2, 3$ for t. Equivalently, we see that $v(s) \geq v(t)$ holds since as shown earlier $X_1 + X_2 + X_4 \geq X_3 + X_2 + X_4 \geq X_3 + X_2 + X_5$. It is convenient to represent the partial order $\succeq$ by omitting the reflexive relations $s \succeq s$ and suppressing any relations that can be derived by transitivity. The result is the *Hasse diagram* [6], shown for this particular example in Figure 1. The largest element in the partial order is $(1, 1, 1, 0, 0)$ and the smallest is $(0, 0, 1, 1, 1)$. The relation $(1, 0, 1, 1, 0) \succeq (0, 1, 0, 1, 1)$ holds because there is a (downward) path from $(1, 0, 1, 1, 0)$ to $(0, 1, 0, 1, 1)$ in the Hasse diagram. However, elements $(1, 0, 0, 1, 1)$ and $(0, 1, 1, 1, 0)$ are *incomparable*: neither $(1, 0, 0, 1, 1) \succeq (0, 1, 1, 1, 0)$ nor $(0, 1, 1, 1, 0) \succeq (1, 0, 0, 1, 1)$ holds. It is also seen that the edges of the Hasse diagram correspond to interchanging an adjacent one-zero pair. This observation holds in general.

LEMMA 2 *$s \succeq t$ is an edge of the Hasse diagram* $\Leftrightarrow$ $s = (\ldots, 1, 0, \ldots)$, $t = (\ldots, 0, 1, \ldots)$.

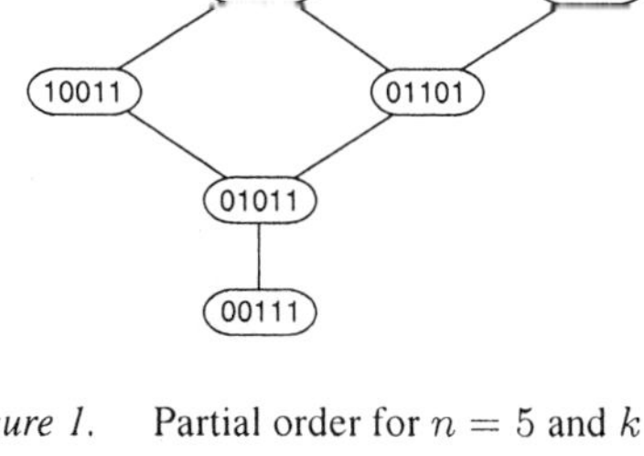

Figure 1. Partial order for $n = 5$ and $k = 3$.

We now use this partial ordering information to describe a state generation algorithm for successively producing the states s in nonincreasing order of $v(s)$. To begin, state $s^0 = (1, 1, \ldots, 1, 0, 0, \ldots, 0)$ has the largest value $v(s)$ among all states. In general, to find the next largest state we maintain a relatively small *candidate set* C of states. At each step of the algorithm, a state s^* having maximum value $v(s)$ among states $s \in C$ is removed from C, and certain *successors* of state s^* are placed into C. This process is repeated until the required number of states have been generated — that is, removed from C. In order for this procedure to generate states s by nonincreasing value $v(s)$, the rule for defining successors needs to be carefully chosen. The following properties of a successor rule will ensure the proper and efficient generation of states using this approach.

(P1) If t is a successor of s then (s, t) is an edge of the Hasse diagram.

(P2) If $t \neq s^0$ then there is a unique state s such that t is a successor of s: that is, s is the unique *predecessor* of t.

Property (P1) ensures that $s \succeq t$ holds; from Lemma 1 we see that $v(s) \geq v(t)$ and so states will be generated in order of nonincreasing value. Property (P2) ensures that every state will appear once (and only once) as a successor of a state. As a result, every state is capable of being generated by the algorithm and duplicate states will not be produced. In addition, Property (P1) ensures that no two successors t_1 and t_2 of a state s are comparable in the partial order (2). If, for example, $t_1 \succeq t_2$ then we would have both $s \succeq t_1$ and $t_1 \succeq t_2$. However, in this case the relation $s \succeq t_2$ could be deduced by transitivity, and so (s, t_2) would not appear as an edge of the Hasse diagram. This contradicts property (P1) of the successor rule, since t_2 is a successor of state s.

A valid successor rule (satisfying the above properties) will then define a spanning tree, rooted at node s^0 in the Hasse diagram. The unique path to a given state s from the root s^0 then defines the unique conditions under which state s will be placed in the set C. Moreover, since successors of a given state are incomparable, it is expected that the size of C will remain manageable, thus avoiding unnecessary comparisons to find at each step the maximum-valued state of C.

We now describe a particular successor rule that satisfies properties (P1) and (P2). Since (P2) is to be satisfied, there must be a unique predecessor s for any state $t \neq s^0$. If $t \neq s^0$ then there is a first occurrence of "0,1" in the state vector t. The unique predecessor of t is thus defined to be that vector s obtained from t by replacing the first occurrence of "0,1" (in a left-to-right scan) by "1,0". Since this is the *first* occurrence of "0,1" in t, the pair "0,1" cannot be preceded by any other "0,1". Consequently, the vector t and its corresponding predecessor s have one of the following four forms:

(a) $t = (0, 1, \ldots)$, $s = (1, 0, \ldots)$;
(b) $t = (1, 1, \ldots, 1, 0, 1, \ldots)$, $s = (1, 1, \ldots, 1, 1, 0, \ldots)$;
(c) $t = (0, 0, \ldots, 0, 0, 1, \ldots)$, $s = (0, 0, \ldots, 0, 1, 0, \ldots)$;
(d) $t = (1, 1, \ldots, 1, 0, 0, \ldots, 0, 0, 1, \ldots)$, $s = (1, 1, \ldots, 1, 0, 0, \ldots, 0, 1, 0, \ldots)$.

Notice that in cases (a) and (b), the first "1,0" in state s can only be preceded (if at all) by a sequence of all "1" entries. On the other hand, for cases (c) and (d), the distinguished "1,0" in state s is preceded by at least one "0". Accordingly, we denote a transition from s to t involving (a) or (b) as a *type 1* transition. A transition involving (c) or (d) is a *type 2* transition.

By reversing the construction in cases (a)-(d) above, we obtain the following successor rule. The two parts of the rule correspond, respectively, to type 1 and type 2 transitions.

(S1) Let the first occurrence of "0" in s be at position $r > 1$. Then the successor t has the same components as s except that $t_{r-1} = 0$, $t_r = 1$.

(S2) Let the first occurrence of "0,1" in s be at positions $r - 1$ and r, with $1 < r < n$. If $s_{r+1} = 0$, then the successor t has the same components as s except that $t_r = 0$, $t_{r+1} = 1$.

Notice that each state s has at most two successors, and so the candidate set C grows by at most one at each step. Moreover the rule (S1)-(S2) defines a valid successor rule. Property (P1) holds by Lemma 2, since each successor involves interchanging an adjacent "1" and "0". Property (P2) holds since the rule is constructed to produce a unique predecessor for each state other than s^0.

To illustrate this successor rule, consider again the case of $n = 5$ and $k = 3$, whose Hasse diagram appears in Figure 1. State $(1,1,0,1,0)$ has two successors: a type 1 successor $(1,0,1,1,0)$ and a type 2 successor $(1,1,0,0,1)$. On the other hand, state $(1,0,1,1,0)$ has a single (type 1) successor $(0,1,1,1,0)$, whereas state $(0,1,0,1,1)$ has a single (type 2) successor $(0,0,1,1,1)$. State $(0,1,1,1,0)$ has no successors at all. Figure 2 displays the edges (s,t) defined by this successor rule. As expected, these edges form a spanning tree rooted at node $s^0 = (1,1,1,0,0)$.

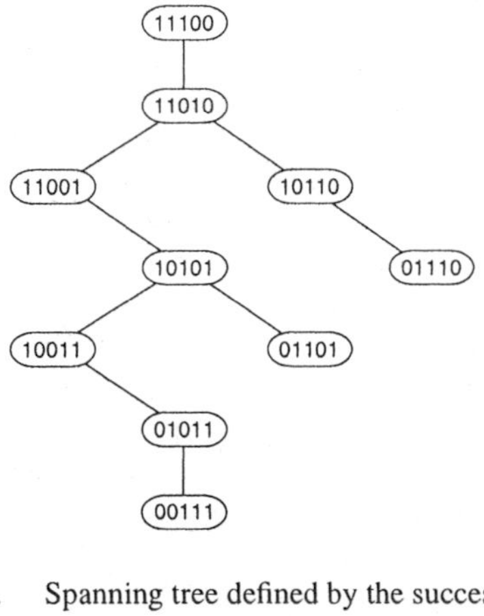

Figure 2. Spanning tree defined by the successor rule.

We present below the proposed algorithm for computing the (upper) tail probability p associated with the specified cutoff value v_o. Since the algorithm successively generates states $\boldsymbol{s}$ by nonincreasing value $v(\boldsymbol{s})$, we obtain as a byproduct the actual distribution values $\Pr(v(I) \geq v)$ for all $v \geq v_o$.

state generation algorithm

Input: n; k; $X_1, X_2, \ldots, X_n$; v_o
Output: tail probability $p = \Pr(v(I) \geq v_o)$
{
  $i := 0$, $\boldsymbol{s}^i := (1, 1, \ldots, 1, 0, 0, \ldots, 0)$, $\mathcal{C} := \emptyset$;
  while $v(\boldsymbol{s}^i) \geq v_o$
  {
    add the successors of $\boldsymbol{s}^i$ to $\mathcal{C}$ using (S1)-(S2);
    $i := i + 1$;
    remove a state $\boldsymbol{s}^i$ with maximum value $v(\boldsymbol{s})$ from $\mathcal{C}$;
  }
  $p := i/\binom{n}{k}$;
}

We illustrate the state generation algorithm using the observed data values $10, 7, 6$ for Group 1 and $8, 6, 4$ for Group 2. Here $n = 6$ and $k = 3$. The mean of the observations in Group 1 is $\overline{X}_1 = \frac{23}{3}$ and the mean for Group 2 is $\overline{X}_2 = 6$. We are then interested in a mean difference $\overline{X}_1 - \overline{X}_2$ at least as extreme as the value $\frac{5}{3}$ observed, or equivalently a value for $\sum_{r=1}^{3} X_{i_r}$ of at least $v_o = 23$.

Ordering the data values gives $X_1 = 10$, $X_2 = 8$, $X_3 = 7$, $X_4 = 6$, $X_5 = 6$, $X_6 = 4$. The algorithm is initialized with $i = 0$, $\boldsymbol{s}^0 = (1, 1, 1, 0, 0, 0)$, and $\mathcal{C} = \emptyset$. The first state generated has value $v(\boldsymbol{s}^0) = X_1 + X_2 + X_3 = 25$. Table 1 shows the progress of the algorithm, which generates five states in order of nondecreasing value until the cutoff value $v_o = 23$ is passed, thus producing the exact value $p = \frac{5}{20} = 0.25$. In this simple example the size $|\mathcal{C}|$ of the candidate set never exceeds four, so selecting a maximum-valued state from $\mathcal{C}$ requires minimal effort.

3. Computational Results

In this section, we discuss certain data structures and implementation details used in our algorithm. We then present some preliminary experimental results to assess the efficacy of our enumeration approach. We have chosen test problems that interpolate between very small problem instances that are easily solved by enumeration and somewhat larger problems for which the Central Limit Theorem can be confidently applied. The computational results indicate

Table 1. Illustration of the state generation algorithm.

Iteration	s^i	$v(s^i)$	*Successor States*^{value}
0	$(1,1,1,0,0,0)$	25	$(1,1,0,1,0,0)^{24}$
1	$(1,1,0,1,0,0)$	24	$(1,0,1,1,0,0)^{23}$, $(1,1,0,0,1,0)^{24}$
2	$(1,1,0,0,1,0)$	24	$(1,0,1,0,1,0)^{23}$, $(1,1,0,0,0,1)^{22}$
3	$(1,0,1,1,0,0)$	23	$(0,1,1,1,0,0)^{21}$
4	$(1,0,1,0,1,0)$	23	$(0,1,1,0,1,0)^{21}$, $(1,0,0,1,1,0)^{22}$
5	$(1,1,0,0,0,1)$	22	

that our algorithm is able to significantly reduce the computational burden of carrying out an enumeration approach.

A closer examination of the example illustrated in Table 1 reveals that some successor states need not be placed in the candidate set. Since we seek to enumerate all states with a value of 23 or greater and the state generation rule produces successors states with nonincreasing values, successor states with a value less than 23 can be *pruned* — i.e., these are not added to the candidate set. There are four such states in this example: $(1,1,0,0,0,1)^{22}$, $(0,1,1,1,0,0)^{21}$, $(0,1,1,0,1,0)^{21}$, and $(1,0,0,1,1,0)^{22}$. These states are not placed in the candidate set, and so the maximum size of the candidate set is actually two rather than four. As will be shown later, a significant number of states can be pruned in practice.

In addition, the state enumeration can be driven by either or both of two different objectives. Either we find all states with a value greater than or equal to a specified cutoff value v_o (thus determining the p-value) or we enumerate sufficient states to find the state value such that the associated tail probability is at least a specified level α. In the former case, the enumeration proceeds as outlined above with the additional benefits of pruning states having value less than the specified cutoff value. Moreover, the algorithm can terminate once the candidate set is empty.

For a specific α level, since the probability of every state is the same, we know precisely how many states must be generated. When the number of enumerated states plus those currently in the candidate set reach this number, the smallest state value found in the candidate set can be used as a nominal cutoff value for pruning subsequent states. Finally, we can use both the cutoff value and the α level when we do not require an exact p-value, but just want to determine if the unknown p-value exceeds the specified α. We might not be able to achieve a tail probability equal to α because the distribution is discrete; moreover, there can be ties occurring among the state values. As will be seen subsequently, such ties also affect the computational requirements of the state generation algorithm.

In our implementation, we chose to use an ordered linked list for the states in the candidate set C. Hence, the candidate state of maximum value is at the beginning of the list and can be selected in constant time. Inserting successor states into the linked list would ordinarily require a linear scan of the list. Since we are also interested in the distribution of the test statistic $v(I)$, we keep a second ordered list of *distinct* state values, which is updated as successors are generated. For each such distinct value, keeping a pointer into the linked list of candidate states makes insertion of a successor only as expensive as finding its value in the second list. If there are few distinct observations, there are many ties in the state values and relatively few distinct state values. As a result, finding the correct place to insert a successor state is fast. Conversely, many distinct observations can lead to many distinct state values and an associated increase in the time required to find the state value and the associated insertion point.

Our approach amounts to a modified bucket sort [1] on the state values in the candidate set. Of course, more sophisticated data structures such as balanced binary trees or k-heaps could be used to maintain the candidate set [1]. For the data we have examined, such methods did not appear warranted.

The problem of determining the tail probabilities of the randomization distribution is essentially combinatorial and exhibits the exponential growth that is characteristic of many such problems. As a baseline, we compare the effectiveness of the state enumeration approach given here with a complete enumeration as given by Edgington [3]. Our objective is to determine the benefits of our approach, as well as its limits of applicability, in the context of *enumeration algorithms*. By contrast, other algorithms appearing in the literature use Monte Carlo methods, asymptotic approximations, characteristic functions, and bounding techniques [5]. The present study does not undertake to compare our enumeration technique with these other, quite different, approaches.

We generated test data from two Weibull distributions, chosen so the t-statistic computed using the true population means and variances was approximately 2. Samples sizes were chosen at an intermediate level — large enough so that a complete enumeration would require significant computation time but small enough so that a traditional t-test would be suspect. The example problems were chosen to have equal sample sizes $n_1 = n_2 = k$, thus maximizing the total number of states for the given total sample size $n = n_1 + n_2$. Table 2 summarizes the computational effort of complete enumeration versus the proposed state generation approach over the range $k = 11, \ldots, 16$.

As expected, the state generation method runs significantly faster than complete enumeration. Both methods were coded in C and executed on a Macintosh PowerPC G5 computer with dual 1.8GHz processors and 512MB of memory. Table 3 gives a more detailed description of the computational effort associated with the state generation method, when a specified cutoff value v_o was used.

Table 2. Comparison of CPU times for complete enumeration vs. state generation.

k	*Number of Required States*	*Complete Enumeration (sec)*	*State Generation (sec)*
11	22,354	0.4	0.1
12	102,983	1.8	0.4
13	529,311	7.5	1.8
14	1,291,463	30.9	4.0
15	5,010,792	126.4	14.5
16	22,702,077	517.0	57.6

Table 3. Computational characteristics of the state generation method.

k	*Total Number of States*	*Number of Required States*	*Number of Pruned States*	*Maximum Candidate Set Size*
11	705,432	22,354	9,403	4,513
12	2,704,156	102,983	38,250	22,217
13	10,400,600	529,311	171,116	107,323
14	40,116,600	1,291,463	368,475	255,258
15	155,117,520	5,010,792	1,427,560	987,095
16	601,080,390	22,702,077	5,267,823	3,697,928

From the results in Table 3, it is evident that a significant number of states can be pruned (not placed in the candidate set) when a cutoff value is given. Also, although the state generation method is considerably faster than complete enumeration and examines only a small fraction of the entire state space, the time savings are achieved at the cost of increased space requirements. In a complete enumeration, the p-value can be determined by simply counting those states with a state value greater than or equal to the cutoff value. This can be done as states are enumerated, involving only a single counter. In our approach, it is necessary to allocate sufficient storage to hold the candidate set. For the example problems studied, the maximum size of the candidate set ranged from 0.6% to 1.0% of the total number of states and was typically 20% of the number of required states. These percentages would be considerably higher if states were not pruned since an average of 1.34 successor states were generated for every enumerated state.

Although our state generation approach achieves a significant reduction in CPU time T compared to complete enumeration, it is clear from Table 2 that the computational effort increases exponentially with k. Indeed an analysis of the model $T = \alpha\beta^k$ was carried out by taking a logarithmic transformation and performing linear regression; the resulting regression line (dashed) is shown in Figure 3. Regression analysis produced a very good fit ($R^2 = 0.996$) and yielded the estimates $\hat{\alpha} = 1.93 \times 10^{-7}$, $\hat{\beta} = 3.37$. Empirically, the CPU time T increases exponentially with k, so we cannot expect that our approach, or any enumeration approach, to be practical for k much larger than 16 (i.e., 32 observations in total). We also note that the space requirements of our approach also appear to increase exponentially. In Figure 4 we observe a near linear log-log plot of the maximum size of $\mathcal{C}$ versus CPU time, using the data presented in Tables 2 and 3. The conclusion is that both space and time requirements are inexorably increasing with k.

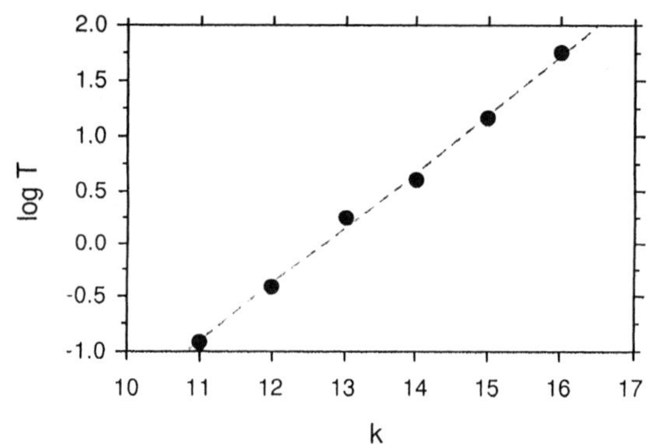

Figure 3. Log T vs. k for the state generation algorithm.

Ties can have a significant impact on the time and space requirements of the state generation approach. For the results in Tables 2 and 3, the sample data values were truncated to three decimal places. This produced an inconsequential number of tied state values compared to using the data at full precision. Table 4 shows the effects of further truncation in the data. Less precision produces more ties in state values and a concomitant increase in the number of states to be enumerated (and the associated p-values). Although the ties result in a larger enumeration, fewer distinct state values mean that less effort is required to find the insertion point for successor states in the state value list. This apparently accounts for the *faster* running times observed as the precision is reduced. As a result, we might expect significantly improved performance of the state generation approach when applied to *ranked data*, in which numerical observations are replaced by their ranks. (In this case, our test procedure reduces to the well-known Wilcoxon rank sum test for comparing two populations.)

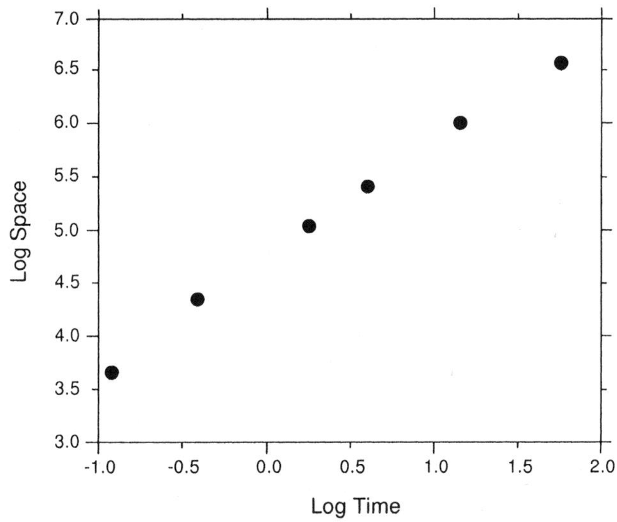

Figure 4. Space and time requirements of the state generation algorithm.

Table 4. Effects of tied state values in the state generation algorithm.

k	*Precision*	*CPU Time (sec)*	*Number of Required States*	*Number of Pruned States*	*Maximum Candidate Set Size*	*p-value*
11	.001	0.1	22,354	9,403	4,513	0.0317
	.005	0.1	22,635	9,507	4,602	0.0321
	.010	0.1	22,752	9,578	4,639	0.0323
12	.001	0.4	102,983	38,250	22,217	0.0381
	.005	0.2	104,380	38,650	22,446	0.0386
	.010	0.1	105,310	39,012	22,706	0.0389
13	.001	1.8	529,311	171,116	107,323	0.0509
	.005	0.6	536,214	172,854	108,214	0.0516
	.010	0.4	541,892	174,693	109,494	0.0521
14	.001	4.0	1,291,463	368,475	255,258	0.0322
	.005	1.2	1,309,525	372,774	259,943	0.0326
	.010	0.9	1,323,557	377,067	263,317	0.0330
15	.001	14.5	5,010,792	1,427,560	987,095	0.0323
	.005	4.7	5,090,388	1,447,104	1,006,708	0.0328
	.010	3.4	5,143,063	1,462,922	1,019,748	0.0332
16	.001	57.6	22,702,077	5,267,823	3,697,928	0.0378
	.005	17.9	23,043,623	5,343,857	3,780,472	0.0383
	.010	12.9	23,198,386	5,394,837	3,809,545	0.0386

As a final observation, the algorithm presented here enumerates states from a single tail of the randomization distribution. By simply negating the observed values (thus reversing their order), states from the other tail can also be generated by our algorithm. Hence, a two-tailed test can be carried out by simply enumerating both extremes of the distribution in turn.

4. Summary and Conclusions

When comparing the means of two groups, the randomization distribution is often a more appropriate statistical model than those associated with conventional random sampling. There is growing use in the scientific community of randomization tests, as well as other resampling methods, which have been widely applied to clinical trials in medicine and biology [7, 8]. Since such tests are computationally intensive, it is important to investigate algorithms that can reduce the computational burden of the combinatorially explosive calculations, especially for smaller sample sizes when the Central Limit Theorem can not be applied with confidence.

We have presented an alternative enumeration method in this paper. Using the algebraic structure of the problem, it is possible to enumerate only the most significant values of the randomization distribution — those in the tails of the distribution. Our approach does require more storage space than complete enumeration, but it runs significantly faster. Indeed, the reduction in computation time can be an order of magnitude and the relative advantage of our approach improves as the number of observations increases. Ultimately, the combinatorial nature of the problem limits enumeration techniques to relatively small sample sizes (here, at most 32 observations in total). However, there are many practical, nontrivial problems with sizes falling within this range that our approach can solve using a reasonable amount of computational resources (time and space). Future research should address how our method compares with the dynamic programming approach of Mehta et al. [10], which forms the basis of the StatXact software package [13].

References

[1] A. V. Aho, J. E. Hopcroft and J. D. Ullman, *The Design and Analysis of Computer Algorithms*, Addison-Wesley, Reading, MA, 1974.

[2] P. Diaconis and S. Holmes, Gray codes for randomization procedures, *Statistics and Computing* **4** (1994), 287–302.

[3] E. S. Edgington, *Randomization Tests*, 3rd ed., Marcel Dekker, New York, 1995.

[4] R. A. Fisher, *Design of Experiments*, Oliver and Boyd, Edinburgh, 1935.

[5] P. Good, *Permutation Tests: A Practical Guide to Resampling Methods for Testing Hypotheses*, 2nd ed., Springer-Verlag, New York, 2000.

[6] R. P. Grimaldi, *Discrete and Combinatorial Mathematics*, 5th ed., Addison-Wesley, Reading, MA, 2004.

[7] J. Ludbrook and H. Dudley, Why permutation tests are superior to t and F tests in biomedical research, *The American Statistician* **52** (1998), 127–132.

[8] B. F. J. Manly, *Randomization, Bootstrap and Monte Carlo Methods in Biology*, 2nd ed., CRC Press, Boca Raton, FL, 1997.

[9] A. W. Marshall and I. Olkin, *Inequalities: Theory of Majorization and Its Applications*, Academic Press, New York, 1979.

[10] C. R. Mehta, N. R. Patel and L. J. Wei, Constructing exact significance tests with restricted randomization rules, *Biometrika* **75** (1988), 295–302.

[11] M. Pagano and D. Tritchler, On obtaining permutation distributions in polynomial time, *Journal of the American Statistical Association* **78** (1983), 435–440.

[12] E. J. G. Pitman, Significance tests which may be applied to samples from any populations, *Journal of the Royal Statistical Society, Series B*, **4** (1937), 119–130.

[13] StatXact software, Cytel Corp., Cambridge, MA. `http://www.cytel.com/`

AN ADAPTIVE ALGORITHM FOR THE OPTIMAL SAMPLE SIZE IN THE NON-STATIONARY DATA-DRIVEN NEWSVENDOR PROBLEM

Gokhan Metan*
Department of Industrial and Systems Engineering, Lehigh University, Bethlehem, PA 18015
gom204@lehigh.edu

Aurélie Thiele†
Department of Industrial and Systems Engineering, Lehigh University, Bethlehem, PA 18015
aurelie.thiele@lehigh.edu

Abstract We investigate the impact of the sample size in the non-stationary newsvendor problem when the underlying demand distribution is not known, and performance is measured by the decision-maker's average regret. The approach we propose is entirely data-driven, in the sense that we do not estimate the probability distribution of the demand and instead rely exclusively on historical data. We propose an iterative algorithm to determine the number of past observations that should be included in the decision-making process, provide insights into the optimal sample size and perform extensive computational experiments.

Keywords: Data-driven optimization, adaptive algorithm, newsvendor problem.

Introduction

In the newsvendor problem, the decision-maker seeks to maximize the profits generated by buying and reselling a perishable product subject to random demand. Ordering too few items will result in dissatisfied customers and lost revenue opportunities; ordering too many will leave the manager with unsold inventory, which must then be salvaged at a loss. Classical applications include magazines, seasonal clothing (e.g., gloves, Halloween costumes), and special

*Research partially supported by the National Science Foundation, grant DMI-0540143.
†Research partially supported by the National Science Foundation, grant DMI-0540143. Corresponding author.

items such as Christmas trees. This problem has been thoroughly investigated to date under a wide range of assumptions such as random yield, multiple products, fixed ordering cost, censored data and unknown demand distribution (see for instance Porteus (2002) for a review of the classical model, Scarf (1958) for an introduction to the distribution-free setting and Gallego and Moon (1993) for extensions). While researchers now acknowledge the difficulty in estimating the probabilities governing the random process, a difficulty first pointed out in Scarf's pioneering work (1958), the distribution-free approaches developed so far build upon some limited probabilistic knowledge such as the first two moments of the demand (Gallego and Moon 1993). It is difficult, however, to obtain such information when demand is non-stationary, as is the case in many practical applications. To address this issue, van Ryzin and McGill (2000) have adapted the stochastic approximation procedure proposed by Robbins and Monro (1951) to the problem of determining optimal seat protection levels in the context of airline revenue management. A key feature of their algorithm is that it does not require any probabilistic information; instead, it adjusts the protection level using optimality conditions derived by Brumelle and McGill (1993). While promising, this approach has had mixed performance in numerical studies (van Ryzin and McGill 2000). Godfrey and Powell (2001) present an adaptive technique based on concave piecewise linear approximations of the value function for the newsvendor problem with censored demand. Recent work by Bertsimas and Thiele (2004) focuses on capturing the decision-maker's risk preferences in the data-driven framework by trimming the number of historical observations. In this approach, demand is stationary and the decision-maker identifies and removes the observations leading to the highest revenue, which depend on the decision variables, without changing the size of the original data set. Bertsimas and Thiele (2004) show that this trimming process can be combined with the optimization procedure in one single step, leading to a tractable linear mathematical formulation, as opposed to a naive iterative algorithm where the decision-maker selects a candidate solution, ranks the revenues generated in the various scenarios, then updates his candidate solution and reiterates. Finally, Levi et. al. (2006) describe an alternative approach to compute sampling-based policies for the newsvendor problem and provide performance guarantees.

In the present paper, we investigate the *impact of the sample size*, i.e., the number of observations included in the decision-making process, on system performance when demand is non-stationary. This differs from trimming, as we keep all data points up to some (to be determined) time period in the past, while the trimming procedure can remove any data point among those in the set; in particular, it can remove recent data and keep older observations. The method we propose is specifically designed to address non-stationary demand distributions and allows insights into the rate of change of the sample size required for good empirical performance. To the best of our knowledge, this is the first attempt to approach the newsvendor problem from a data-driven perspective through the *dynamic update* of the number of historical data points considered. The promise of the methodology lies in the decision-maker's ability to react quickly to observations without making estimation errors.

The remainder of the paper is structured as follows. We formulate the problem in Section 1, and present preliminary results on the behavior of the average regret in Section 2. In Section 3, we propose and test a distribution-free algorithm to determine the optimal sample size. Section 4 contains concluding remarks.

1. Problem Setup

In this section, we formulate the model in mathematical terms and describe the data-driven approach. We consider the classical newsvendor problem (Porteus 2002) with the following notation:

c: the unit ordering cost,
p: the unit price,
s: the salvage value,
x: the order quantity,
D: the random demand.

Specifically, the newsvendor orders an amount x at unit ordering cost c before knowing the value d taken by the demand. Then demand is realized, and the newsvendor sells $\min(x, d)$ newspapers at unit price p. At the end of the time period, he salvages the remaining inventory $\max(0, x - d)$ at salvage price s.

The newsvendor's profit for a given demand realization d is given by:

$$\begin{aligned} \pi(x,d) &= -c\,x + p\min(x,d) + s\max(0,x-d), \\ &= (p-c)\,x - (p-s)\max(0,x-d). \end{aligned}$$

For this specific instance, it would have been optimal for the decision-maker to order exactly the demand d, thus realizing the profit:

$$\pi^*(d) = (p-c)\,d.$$

The difference between these two terms, $\pi^*(d) - \pi(x,d)$, is called the *regret* of the decision-maker and is always nonnegative. The problem of minimizing the newsvendor's expected regret for a given demand distribution is then formulated as:

$$\min_{x \geq 0} (p-c)\,(E[D]-x) + (p-s)\,E\max(0,x-D), \tag{1}$$

where $E[D]$ is the expected demand. Since $E[D]$ is a constant, Problem (1) is equivalent to maximizing the expected revenue. If the decision-maker has N historical data points $d_1, \ldots, d_N$ at his disposal and seeks to minimize the sample average regret over these points, Problem (1) becomes:

$$\min_{x \geq 0} (p-c)(\overline{d}-x) + \frac{(p-s)}{N} \sum_{i=1}^{N} \max(0, x-d_i), \tag{2}$$

with $\overline{d}$ the sample average demand.

THEOREM 1.1 (OPTIMAL ORDER IN DATA-DRIVEN PROBLEM) *The optimal solution of Problem (2) is given by:*

$$x^* = d_{<j>} \text{ with } j = \left\lceil \frac{p-c}{p-s} N \right\rceil,$$

where $d_{<.>}$ is the ordered data set such that $d_{<1>} \leq d_{<2>} \leq \ldots \leq d_{<N>}$.

Proof: Follows from studying the sign of the slope of the objective. □

Remark: When the underlying probabilities are known exactly, and the demand has a continuous cumulative distribution function F, the optimal order is given by $F^{-1}((p-c)/(p-s))$ (Porteus 2002), which matches the asymptotic value of the order in Theorem 1.1 when $N \to \infty$.

The quality of the historical data used, i.e., the relevance of past data points to predict future demand, is obviously critical to the performance of the approach. In this work, we do not perform any forecasting on the future demand and instead investigate the impact of the number of historical observations as a control variable to improve system performance. When the underlying demand distribution is stationary, it is of course optimal to use all the historical observations available. In the presence of seasonality factors or other non-stationary trends, however, past data points become obsolete or irrelevant as the time since these observations were made increases. Under such circumstances, it is more appropriate to focus solely on the most recent data. The goal of this work is to quantify how recent these observations should be in order to minimize the decision-maker's sample regret. In Section 2 we provide some insights into this question by performing numerical experiments for various time-varying demand distributions.

2. Preliminary Results

In this section, we describe and analyze preliminary numerical experiments, which highlight the importance of the data set size to achieve good system performance. Table 1 summarizes the characteristics of the demand distributions used in the simulations. Specifically, we consider Gaussian random variables with time-varying parameters, where we change: (i) the mean alone (Section 2.1), (ii) the standard deviation alone (Section 2.2), and (iii) both the mean and the standard deviation (Section 2.3). We chose the numerical values to illustrate a wide range of settings, e.g., slowly-increasing mean with constant variance, constant coefficient of variation, non-monotonicity. In Section 2.4, we present another set of experiments, where we investigate seasonality factors by incorporating sinusoidal trends in the mean of the distribution. We generate historical data sets of sizes ranging from 1 to 300 for 5000 periods of iteration run length and consider 20 independent iterations. We set the parameter values of the newsvendor problem to: $p = 10,\ c = 7,\ s = 5$.

Table 1. Demand processes used in the experiments.

Type of non-stationarity	*Demand distribution* $N(\mu_t, \sigma_t)$
Time-varying μ	$N(20 + t^{1/3}, 7)$, $N(20\,t^{1/5}, 7)$, $N(20\,t^{1/4}, 7)$, $N(20\,t^{1/3}, 7)$, $N(20\,t^{1/2}, 7)$, $N(50\ \ln t^2, 17)$, $N(\max\{40, 40\,\lvert t-2500\rvert^{1/4}\}, 17)$,
Time-varying σ	$N(60, 2 + t^{1/3})$, $N(60, 2\,t^{1/3})$, $N(60, t^{1/5})$, $N(50, \ln t^2)$, $N(40, \max\{7, 7\,\lvert t-2500\rvert^{1/4}\})$,
Time-varying μ and σ	$N(3\,t^{1/3}, t^{1/3})$, $N(5\,t^{1/4}, t^{1/3})$, $N(5\,t^{1/6}, t^{1/3})$, $N(50\ \ln t^2, \ln t^2)$, $N(\max\{40, 40\,\lvert t-2500\rvert^{1/4}\}, \max\{7, 7\,\lvert t-2500\rvert^{1/4}\})$.

2.1 Mean nondecreasing with time

We investigate here the impact of the number of historical data points when the mean of the demand distribution varies, and in particular is nondecreasing with time. When there is no uncertainty, i.e., the demand is a deterministic time series, the regret is minimized by considering only the most recent data point ($N = 1$), as that point will be the closest among past realizations to the demand in the next time period. At the other extreme, when demand is stationary it is optimal to keep all observations. We note that the marginal impact of information decreases once the sample is large enough, as collecting additional data points will not change the order implemented significantly.

We now quantify this insight in mathematical terms. For convenience, we denote by r the ratio $(p-c)/(p-s)$. Since the average regret is continuous in the order, we focus on the optimal order $F^{-1}(r)$, where F is the cumulative distribution of the demand, and actual order $d_{<\lceil rN \rceil>}$, rather than on the regret itself. The following lemma reviews useful results on the distribution of the sample quantile when the demand distribution is continuous (as is the case here) with probability density function f:

LEMMA 2.1 (FERGUSON 1996)
(i) $\sqrt{N}\left(d_{<\lceil rN \rceil>} - F^{-1}(r)\right)$ *is asymptotically Normally distributed with mean 0 and standard deviation* $\sqrt{r(1-r)}/f(F^{-1}(r))$.

(ii) $P(|d_{<\lceil rN\rceil>} - F^{-1}(r)| > \epsilon) \rightarrow 0$ *exponentially fast in* N. *Specifically,*

$$P(|d_{<\lceil rN\rceil>} - F^{-1}(r)| > \epsilon) \leq 2\, e^{-2\,N\,\delta_\epsilon^2},$$

with $\delta_\epsilon = \min\{F\left(F^{-1}(r) + \epsilon\right) - r, r - F\left(F^{-1}(r) - \epsilon\right)\}$.

We now investigate the impact of non-stationarity on the optimal sample size. To motivate the analysis, we first present the numerical results for five demand processes, which are all Normally distributed with standard deviation 7. The processes are distinguished by their mean, which increases more and more with time. Specifically, we consider five cases for the mean at time t: (i) $\mu_t = 20 + t^{1/3}$, (ii) $\mu_t = 20\,t^{1/5}$, (iii) $\mu_t = 20\,t^{1/4}$, (iv) $\mu_t = 20\,t^{1/3}$, (v) $\mu_t = 20\,t^{1/2}$. Due to space constraints we do not plot the results for cases (i), (iii) and (v). When the increase in the mean of the distribution is slow (cases (i)-(iii), see Figure 1 for case (ii)), the sample regret exhibits the same qualitative behavior as for a stationary distribution and using large numbers of historical data points results in low regret values. (The curve does exhibit a very slight trend upward.)

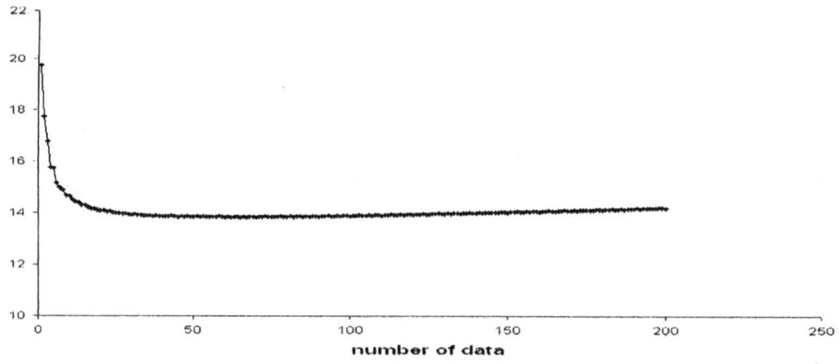

Figure 1. Average regret as a function of sample size for $N(20\,t^{1/5}, 7)$.

When the mean of the demand distribution increases faster (cases (iv) and (v), see Figure 2 for case (iv)), we observe a *trade-off* between keeping only the most recent data points (because the distributions are closest to the ones in the future) and using many observations to smooth out randomness. This results in a decrease in the average regret up to some data set size N^*, followed by an increase when the data set size grows beyond that threshold value. Furthermore, the optimal sample size N^* decreases towards 1 as the rate of variation in the mean increases. Intuitively, the benefit of smoothing the randomness is cancelled out by the fact that the data points belong to very different distributions. This trade-off between too many and too few data points in the sample can be

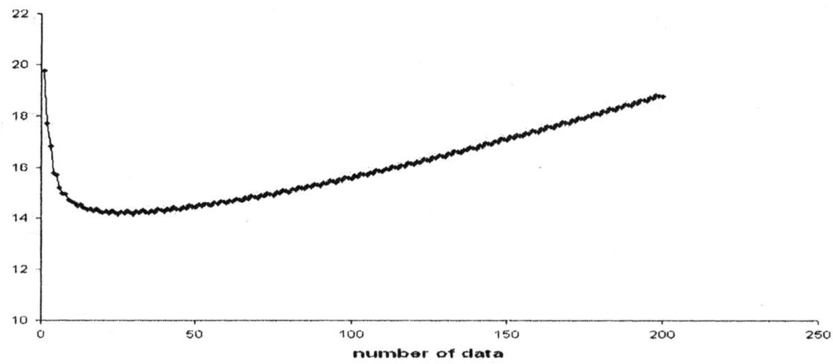

Figure 2. Average regret as a function of sample size for $N(20\,t^{1/3}, 7)$.

characterized in mathematical terms when variability of the demand is small and the distribution is Gaussian. We represent the demand observations d_t as: $d_t = \mu_t + \sigma\, z_t$, where z_t are observations from the standard Normal random variable. d^* denotes the demand at the next time period, i.e., the optimal order, with $d^* = \mu^* + \sigma\, z^*$.

THEOREM 2.2 (NON-STATIONARY DEMAND)
Assume that demand variability is small, so that $d_{<j>} = \mu_{<j>} + \sigma\, z_{<j>}$ for all $j = 1, \ldots, N$.
(i) The difference between the actual and optimal orders, $d_{<\lceil rN\rceil>} - d^*$, asymptotically obeys a Normal distribution with mean $\mu_{<\lceil rN\rceil>} - \mu^*$ and standard deviation $\frac{\sigma}{f(F^{-1}(r))}\sqrt{\frac{r(1-r)}{N}}$.
(ii) If $d_{<\lceil rN\rceil>} - d^*$ obeys the distribution in (i), the average regret given N data points, $E[R_N]$, can be written as:

$$E[R_N] = (p-c)\left(\mu^* - \mu_{<\lceil rN\rceil>}\right) + (p-s)\, E\max\left(0, \mu_{<\lceil rN\rceil>} - \mu^* + \frac{\sigma}{f(F^{-1}(r))}\sqrt{\frac{r(1-r)}{N}}\, Z\right),$$

where Z is the standard Gaussian random variable. Note that, since the mean demand is increasing, $\mu^* > \mu_{<\lceil rN\rceil>}$ and $E[R_N] \approx (p-c)\left(\mu^* - \mu_{<\lceil rN\rceil>}\right)$ when N becomes large.
(iii) To decrease average regret by adding one more data point to a sample of size N, it is sufficient to have:

$$(\mu^* - \mu_{<\lceil rN\rceil>})\sqrt{N} < \frac{\sigma\sqrt{r(1-r)}}{f(F^{-1}(r))}\,\Phi^{-1}(1-r). \quad (3)$$

Proof: (i) and (ii) follow immediately from Lemma 2.1 applied to the observations of the standard Normal random variable.
To obtain (iii), we note that, with $a = \frac{\sigma\sqrt{r(1-r)}}{f(F^{-1}(r))}$ and $b_N = \mu_{<\lceil rN\rceil>} - \mu^*$:

$$E[R_{N+1}] - E[R_N] = (p-c)\ (b_N - b_{N+1}) \\ +(p-s)\ \left[E\max\left(0, b_{N+1} + \tfrac{a}{\sqrt{N+1}}Z\right) - E\max\left(0, b_N + \tfrac{a}{\sqrt{N}}Z\right)\right].$$

Hence,

$$\begin{aligned} E[R_{N+1}] - E[R_N] &\le (p-s)\{r\,(b_N - b_{N+1}) \\ &\quad +E[\{(b_{N+1} - b_N) + a(\tfrac{1}{\sqrt{N+1}} - \tfrac{1}{\sqrt{N}})Z\}1_{\{Z\ge -\frac{b_N\sqrt{N}}{a}\}}]\} \\ &\le (p-s)\,(b_N - b_{N+1})\left[r - P\left(Z \ge -\tfrac{b_N\sqrt{N}}{a}\right)\right], \end{aligned}$$

where we have used the fact that the mean demand increases with time and additional data points are the least recent (hence the smallest under our assumption of low variability) to show that: $0 > b_N \ge b_{N+1}$ and $-b_N\sqrt{N}$ is a positive number increasing with N. We conclude by studying the sign of $r - P\left(Z \ge -\frac{b_N\sqrt{N}}{a}\right)$. □

Remark: Equation (3) suggests that the smallest N such that:

$$(\mu^* - \mu_{<\lceil rN\rceil>})\sqrt{N} \ge \frac{\sigma\sqrt{r(1-r)}}{f(F^{-1}(r))}\,\Phi^{-1}(1-r)$$

will achieve a good balance between keeping only a few data points to take advantage of the non-stationarity of the demand and using as many as possible to smooth out the effects of randomness. It also highlights the importance of having a small difference between μ^* and $\mu_{<\lceil rN\rceil>}$; no matter how small N is, if this difference increases sufficiently, the sample size N will become too large to yield good practical performance. This coincides with the observations drawn from Figure 2.

In Figure 3 we summarize these insights by plotting the average regret for three different demand functions: *deterministic increasing demand* (DD), *stationary stochastic demand* (SSD), and *non-stationary stochastic demand* (NSD). Specifically, for the deterministic increasing demand we use $D_t = 20\,t^{1/3}$, for the stationary stochastic demand we use $D_t \sim N(20, 7)$, and for the non-stationary stochastic demand we use $D_t \sim N(20\,t^{1/3}, 7)$.
From Figure 3, we see that NSD behaves like SSD for small historical data set sizes until N reaches a threshold N^*, which achieves the minimum average regret. The similarity between NSD and SSD is explained by the fact that the

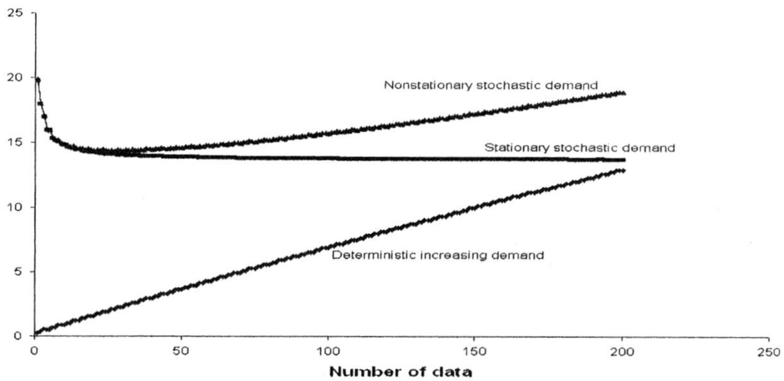

Figure 3. Behavior of regret for three types of demand characterizations.

effect of variability in the parameters was not captured in such a small sample. For $N > N^*$, however, NSD starts behaving like DD and the average regret deteriorates as N increases. Because we keep more data than necessary and the average demand is increasing, we find ourselves in a situation where we consistently under-order, i.e., order less than the actual demand, and using more data points accentuates this trend.

From these experiments, we conclude that the mean of a non-stationary demand process has a significant impact on the average regret, and the sample size should be chosen carefully to reflect the importance of this parameter. In the examples presented in this section, it was optimal to use between 1 and 50 data points. In Section 3 we propose an adaptive algorithm to determine the optimal sample size.

2.2 Variance increasing with time

In this set of experiments we keep the mean of the demand distribution constant and examine the behavior of the average regret when the standard deviation varies with time. Specifically, we consider the following three functions: (i) $\sigma_t = 2\,t^{1/3}$, (ii) $\sigma_t = 2 + t^{1/3}$, and (iii) $\sigma_t = t^{1/5}$. Figure 4 shows the corresponding average simulated regrets with respect to the sample size.
These empirical results suggest that the average regret decreases as the sample size increases up to some point N^*, and then stabilizes. Therefore, using any number of historical data points greater than N^* in the decision process is optimal. The main reason for this behavior is that the optimal order is now given by: $x^* = \mu + (\sigma.\,z.)_{<\lceil rN \rceil>}$, and even when variability is small it is not possible to rewrite $(\sigma.\,z.)_{<\lceil rN \rceil>}$ as $\sigma_{<\lceil rN \rceil>}\, z_{<\lceil rN \rceil>}$. In particular, since the observa-

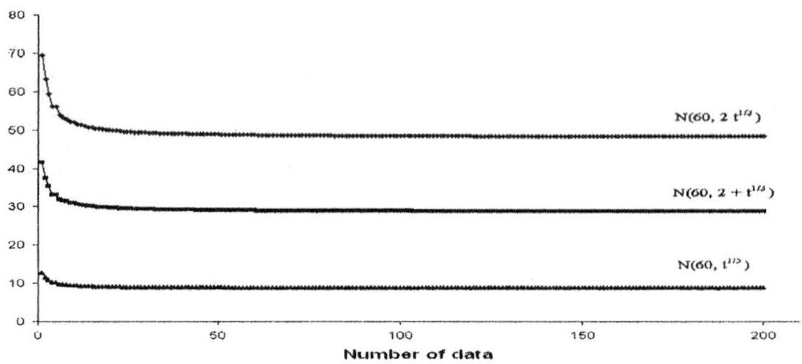

Figure 4. Average regret as a function of sample size.

tions z_j, $j = 1, \ldots, N$, are drawn from a standard Normal distribution, they are as likely to be positive as negative, which will obviously alter the ranking of the data.

Another observation is that, when the change in the standard deviation is slow, the average regret stabilizes at smaller N^* values. For instance, the average regret reaches its minimum after approximately $N^* = 8$ data points for $\sigma_t = t^{1/5}$, and after approximately $N^* = 40$ data points for $\sigma_t = 2\,t^{1/3}$. We also note that when the rate of change in the variance increases, regret increases. This corresponds to an upward shift of the plots in Figure 4.

2.3 Mean and Variance increasing with time

In this section we consider demand distributions for which mean and variance vary simultaneously. We perform three sets of experiments, the parameters of which are provided in the last part of Table 1. In the first set of experiments, we investigate the behavior of the average regret under different *coefficients of variation* (constant, slowly increasing, fast increasing). Figure 5 shows the average regret as a function of the number of data points. We observe that the behavior of the response functions, i.e., the sample regret, is similar to the behavior exhibited in the stationary stochastic demand case. In other words, the regret decreases as the sample size increases, and stabilizes after a threshold size N^*. This is in particular true when both mean and standard deviation increase in $t^{1/3}$. Recall from Figure 2 that, when the mean increases in $t^{1/3}$ but the standard deviation is kept constant, there exists a unique optimal sample size N^* and increasing the data set past that threshold increases the average regret. Hence, this suggests that a standard deviation increasing with time (as the mean) might contribute towards stabilizing the average regret, i.e., increasing the range of near-optimal sample sizes.

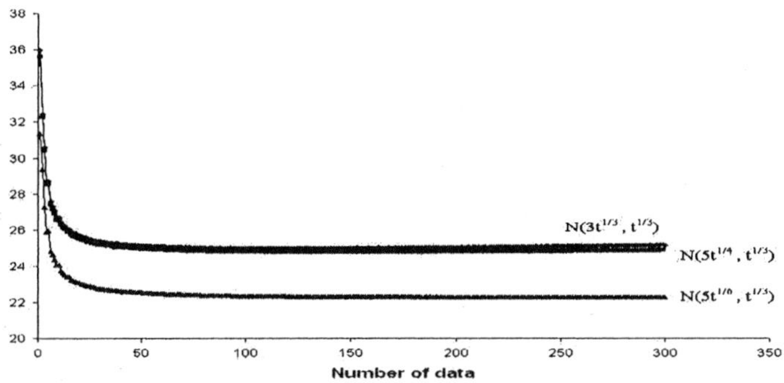

Figure 5. Average regret as a function of sample size in Experiment 1.

In the second set of experiments (see Figure 6), we use logarithmic functions for mean and variance of the demand distribution to gain more insights into the role of the rate of increase. Again, we observe that (i) when the mean is constant the behavior of the average regret is SSD–like, where any sample size past a threshold will yield optimal results, and (ii) having a constant coefficient of variation rather than a constant standard deviation decreases the average regret. In particular, the performance under demand distribution $N(50\ \ln t^2, 17)$ is worse than the performance under $N(50\ \ln t^2\ , \ln t^2)$.

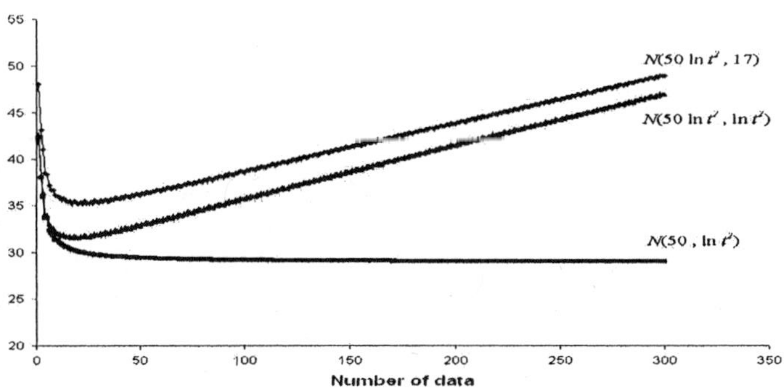

Figure 6. Average regret as a function of sample size in Experiment 2.

In the third set of experiments (see Figure 7), we use piecewise linearly decreasing and then increasing functions for the mean and variance to study the impact of monotonicity on the results. (Figure 7 only shows the results under the demand distribution $N(\max\{40, 40\,|t-2500|^{1/4}\}, 17)$. The results under other two demand functions are very similar to that shown in Figure 6; therefore we do not provide them here.) Again we observe SSD–like behavior, motivated by the slow increase in the mean of the demand distribution. Hence, using data set sizes larger than a threshold value is sufficient to ensure good performance.

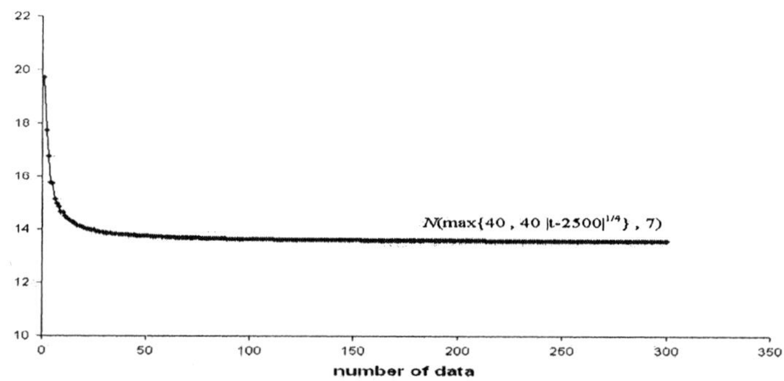

Figure 7. Average regret as a function of sample size in Experiment 3.

2.4 Sinusoidals

In order to test the effect of seasonality on the problem, we perform additional experiments when the mean demand is sinusoidal. (We do not present experiments about sinusoidal standard deviations here as the average regret exhibits, as before, SSD–like behavior in this case.) We consider the following means with a constant standard deviation equal to 7:

$$\begin{array}{rcll} \mu_t^1 &=& 200 + 5 \sin \frac{2\pi t}{20} & : \text{ small amplitude, high frequency} \\ \mu_t^2 &=& 200 + 180 \sin \frac{2\pi t}{20} & : \text{ large amplitude, high frequency} \\ \mu_t^3 &=& 200 + 5 \sin \frac{2\pi t}{500} & : \text{ small amplitude, low frequency} \\ \mu_t^4 &=& 200 + 180 \sin \frac{2\pi t}{500} & : \text{ large amplitude, low frequency} \end{array}$$

The first two mean parameters (μ_t^1 and μ_t^2) as well as the last two (μ_t^3 and μ_t^4) are similar in terms of their frequencies. Also, the first and third, and the second and fourth mean parameters are similar in terms of their amplitudes. The simulation results for these four experimental conditions are presented in Figures 8 and 9. We first compare the results given for the two amplitude levels: small on the left panel of Figures 8 and 9 and large on the right panel. When amplitude is small, regardless of the frequency level, average regret decreases when we increase the number of data points from $N = 1$; we observe the opposite behavior when amplitude is large. We explain this behavior as follows. When amplitude is small, the impact of seasonality is weak; therefore, using small sample sizes deteriorates performance, since we lose information on the variance. On the

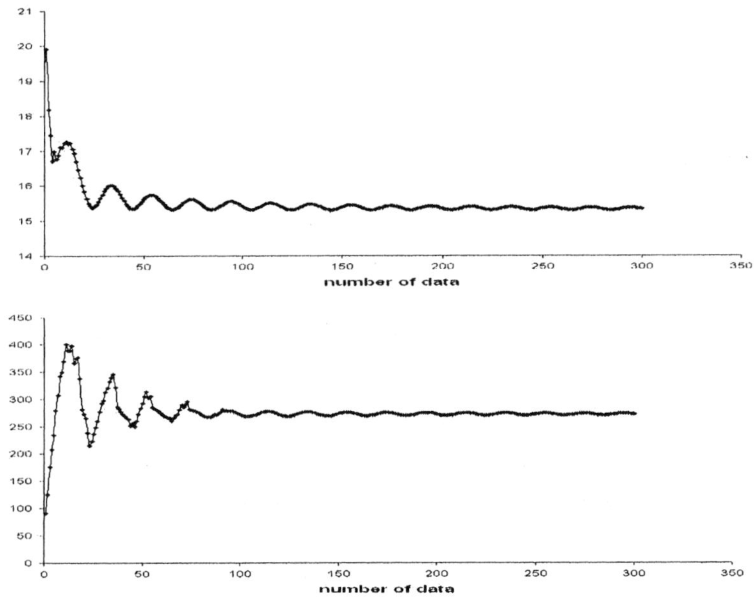

Figure 8. Results for high frequency: demand distribution $N(200 + 5 \sin \frac{2\pi t}{20}, 7)$ (top) and $N(200 + 180 \sin \frac{2\pi t}{20}, 7)$ (bottom).

other hand, when the amplitude of the seasonality effect is large, it is beneficial to use the most up-to-date demand information, which in turn is possible only if we use small historical data sets.

We now consider the impact of seasonality frequency. If we compare the results in Figure 8 with those in Figure 9, we can see that the response function is smoother when frequency is low, but when frequency is high, the average regret is extremely sensitive to the data set size and exhibits up and down peaks. This sensitivity decreases when we increase the data set size (see Figures 10 and 11), in which case we observe the convergence of the average regret to a near-optimal value. We explain this behavior as follows. When the historical data set size is kept small, we utilize the most recent data points. This set, however, might represent the low-demand season when we switch to the high-demand season. This might cause a time lag between the demand and the sample demand data used, which would result in poor performance (see Figure 10). Therefore, using small historical data sets in high frequency seasonal demand environments is fraught with risks.

On the other hand, when the historical data set is kept large, we have the opportunity to recover a significant part of the seasonality and can utilize this

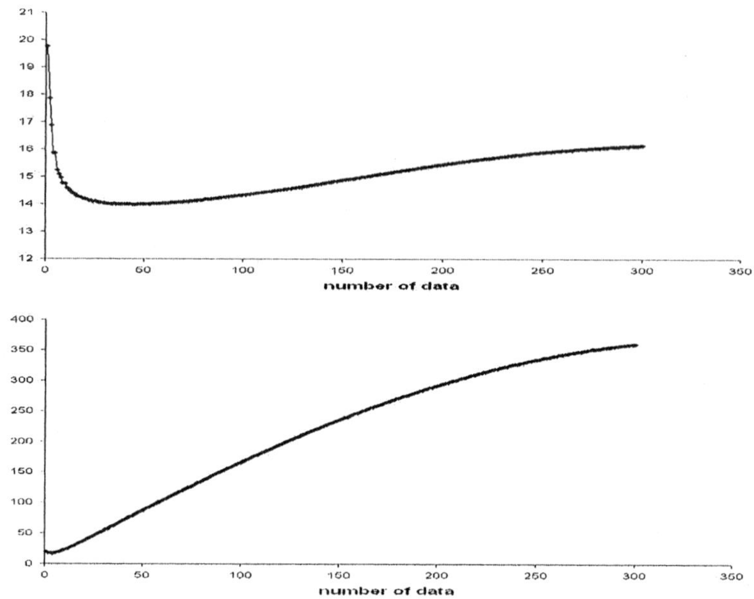

Figure 9. Results for low frequency: demand distribution $N(200 + 5 \sin \frac{2\pi t}{500}, 7)$ (top) and $N(200 + 180 \sin \frac{2\pi t}{500}, 7)$ (bottom).

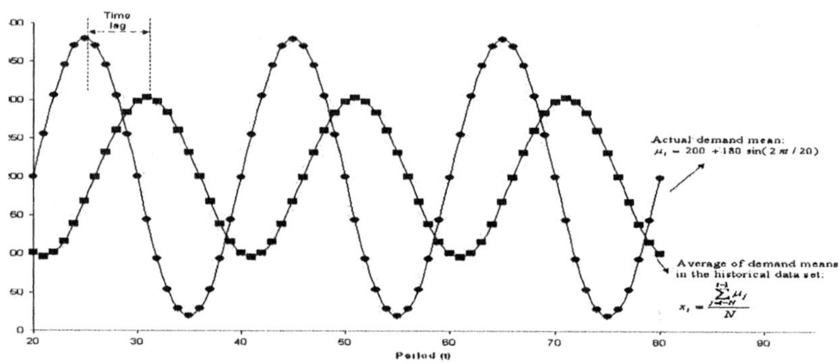

Figure 10. Behavior of actual demand mean in time and the average of sample mean for data set size $N = 11$.

information for better demand estimates in the future. A disadvantage is that, since we have vast amounts of historical data at our disposal, we lose the most recent information about the state of the seasonality. In other words, since we treat all data points in our sample in the same fashion, we will not draw enough benefit from the most recent observations, and the influence of new observations on a large historical data set is imperceptible (see Figure 11).

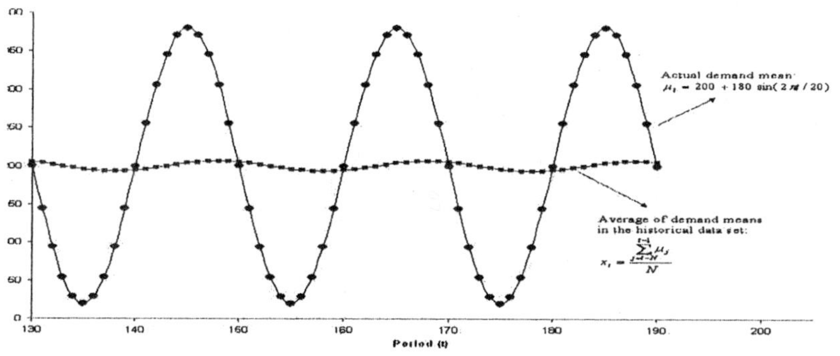

Figure 11. Behavior of actual demand mean in time and the average of sample mean for data set size $N = 125$.

3. An Adaptive Algorithm

In Section 2 we investigated the behavior of the average regret under several demand processes as a function of the size of the data set. We now propose an algorithm to determine the appropriate value of this sample size and test the performance of the approach for the demand distributions used in Section 2. We also provide insights into the fine-tuning of the key parameter in the algorithm.

We assume that we have no a-priori information about the distribution of the demand. Our objective is to develop an algorithm that successively updates the size of the data set N in order to achieve smaller regret values. We also want the convergence towards the terminal value to occur quickly when demand is stationary or slowly changing. (Convergence for non-stationary demand processes has little meaning, since the decision-maker does not know the future demand.) At a high level, the algorithm builds upon the following observations:

- If the current value of N has led us to underestimate the demand, we make adjustments to reduce the gap between the actual demand and the order implemented. Similarly, if we have overestimated the demand, we make adjustments in the opposite direction.
- The scope of the adjustment (in absolute value) is the same whether we underestimate or overestimate the demand, i.e., we consider both outcomes to be equally undesirable and penalize them equally.
- The extent of the adjustment depends on the penalty function. Linear penalties, where the change in the sample size is proportional to the esti-

mation error, seem to yield a good trade-off between speed of convergence and protection against numerical instability.

Once we approach the optimal N^* value, the estimation error gets smaller and the convergence occurs. The pseudocode of the algorithm is given below.

Algorithm. (Reactive Data Set Size Update Rule)
$k \leftarrow 1,\ S \leftarrow S_0,\ N \leftarrow N_0 = |S_0|$
repeat
 Compute order quantity $x_k^* = D_{<j>} \in S, j = \left\lceil \frac{p-c}{p-s} N \right\rceil$;
 Observe the current period's demand D_k;
 Assign $N \leftarrow \max\left\{1, \left\lfloor N - \gamma 100 \left(\frac{x_k^* - D_k}{x_k^*}\right)\right\rfloor\right\}$;
 Update S to have N most recent observations;
 $k \leftarrow k + 1$;
 if $k \geq PHL$, the Planning Horizon Length
 set $N_{final} \leftarrow N$ and $N_{average}$;
end(repeat)

In the algorithm, k and S represent the time period and the historical data set used, respectively. The γ parameter of the algorithm is used as a scaling factor which affects the estimation error expressed in percents. This parameter plays a critical role in the performance of the algorithm, as observed in the experiments below. We consider a range of $[-4, 4]$ for γ with 0.05 increments and use the same demand functions as in Section 2.4, as well as the stationary distribution. In all the experiments, the initial historical data set size is taken to be 9 ($N_0 = |S_0| = 9$).

We now present the results when demand is stationary. In Figure 12, the ratio r is equal to $\frac{10-7}{10-5} = \frac{3}{5} > \frac{1}{2}$, and we observe that $\gamma < 0$ is optimal. This

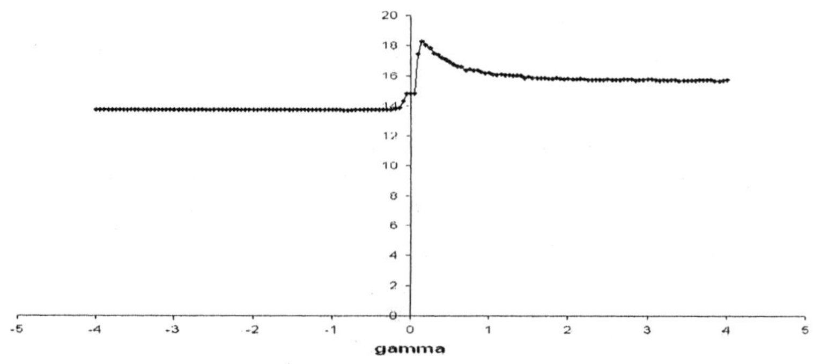

Figure 12. Regret as a function of γ for stationary distribution $N(200, 7)$ with $r > 1/2$.

result holds much more generally, as explained in Theorem 3.1.

THEOREM 3.1 (γ IN STATIONARY CASE)
(i) The expected change in the data set size at the next iteration $E[\Delta N]$ is proportional to γ and to $E[D] - x^*$, where x^* is the optimal order computed with the present data of size N. Specifically:

$$E[\Delta N] = \frac{100\,\gamma}{x^*}(E[D] - x^*).$$

(ii) If $r > \frac{1}{2}$, it is optimal to take $\gamma < 0$, provided that N exceeds a threshold, in the sense that on average the size of the data set increases.
(iii) Similarly, if $r < \frac{1}{2}$, it is optimal to take $\gamma > 0$, provided that N exceeds a threshold.

Proof: (i) We have:

$$\begin{aligned} E[\Delta\, N] &= -\gamma\frac{100}{x^*}\,E\left(|x^* - D|\,1_{\{D\leq x^*\}}\right) + \gamma\frac{100}{x^*}\,E\left(|x^* - D|\,1_{\{D>x^*\}}\right) \\ &= \frac{100\,\gamma}{x^*}\,E\left([D - x^*]\,1_{\{D>x^*\}} - [x^* - D]\,1_{\{D\leq x^*\}}\right) \\ &= \frac{100\,\gamma}{x^*}\,(E[D] - x^*)\,. \end{aligned}$$

(ii) If $r > \frac{1}{2}$, $E[D] < x^*$ with high probability provided that N exceeds a threshold (this threshold can be computed using Lemma 2.1, since $x^* = d_{<rN>}$). The proof of (iii) is similar. □

Figures 13 and 14 show the performance and average N values achieved for various values of γ when the average demand is sinusoidal with high frequency. (Due to space constraints, we refer the reader to Metan and Thiele (2006) for the case with low frequency.) When the γ parameter is well chosen, the algorithm converges to the optimal value of N in three out of four cases, specifically, low frequency with any amplitude and high frequency with small amplitude (see Figure 13). In the remaining case (high frequency and high amplitude; see Figure 14), the best-regret value produced by the algorithm is twice the optimal value. This is the most difficult case, however, since both the amplitude and the frequency of the seasonality are high. Thus, it is difficult for the algorithm to react fast enough to ensure that the value of N converges to near-optimal values.

These results emphasize that the parameter γ, and in particular its sign, must be chosen carefully. In Figure 13, which depicts the case with small amplitude, high frequency, we observe that any $\gamma \leq -0.3$ will yield good practical performance, but a positive value of γ will drastically increase the regret. Qualitatively, this indicates that we should *decrease* the sample size, i.e., focus on the most recent data, when the actual demand is greater than the order.

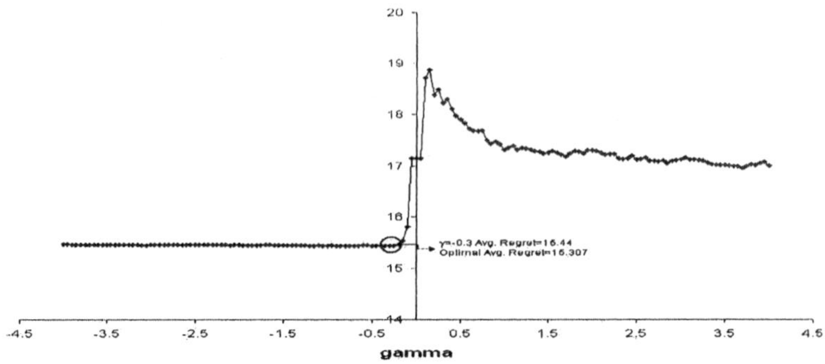

Figure 13. Regret as a function of γ for demand distribution $N(200 + 5 \sin \frac{2\pi t}{20}, 7)$.

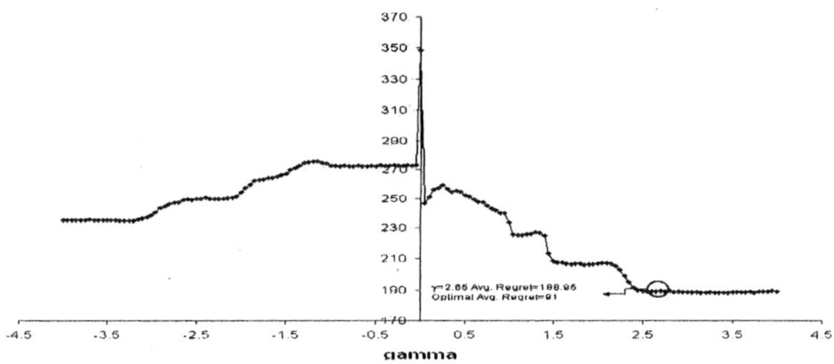

Figure 14. Regret as a function of γ for demand distribution $N(200 + 180 \sin \frac{2\pi t}{20}, 7)$.

When the actual demand is smaller, we incorporate older data points. Figure 14 indicates that, for the high-amplitude, high-frequency case, it is optimal to take $\gamma \geq 2.5$; we also note that, once the sign of γ has been chosen, it is optimal to take $|\gamma|$ large, i.e., make "big" updates at each step (for instance if $|\gamma| = 3$, a difference between the order and the actual demand of 5% will change the data set by 15 points.) Increasing the amplitude of a sinusoidal demand function thus brings a significant change to the optimal value of γ. Metan and Thiele (2006) investigate this point in further detail by considering other amplitude values; in particular, the algorithm appears to be robust to the choice of the γ parameter.

In summary, the numerical experiments suggest that: (i) the algorithm exhibits promising empirical behavior when the scaling parameter γ is equal to its optimal value, (ii) there is no one-size-fits-all value of γ that would be optimal in all circumstances; in particular, whether the ratio $(p - c)/(p - s)$ is above or below $1/2$ plays an important role in selecting this value, (iii) good performance requires the fine-tuning of γ, with a particular focus on very small and very large values of positive and negative sign, (iv) fine-tuning can be done by keeping track of several orders (the actual one and the ones derived with other γ parameters) and adjusting γ when one value consistently over-performs the others.

Hence, updating the sample size using piecewise linear decision rules (each piece corresponding to a value of γ) appears to be the most promising choice for this adaptive algorithm. Figure 15 depicts the evolution of γ when such an algorithm is implemented for a Normal demand distribution with mean 200 and standard deviation $5\,sin(2\pi\,t/500)$ and four values of γ are available: -3.0, -0.1, 0.1 and 3.0. (These values were chosen based on the behavior of the regret function observed in the previous experiments: the left and right tails of the regret appear to stabilize around -3.0 and 3.0, respectively, and there are peaks in the regret for values of γ close to zero, e.g., -0.1 and 0.1.) Throughout the simulation run, the algorithm implements the value of the data set given by the update rule computed with the active value of γ, but also keeps track of the regret that would have been achieved for the other, non-active, values of γ. Every 10 time periods, the algorithm reconsiders its choice of γ; if a non-active value has performed better than the one currently in use, in the sense that it yields a smaller average regret where the average is computed over the last 10 time periods, the algorithm will change values.

Figure 15. Evolution of γ as a function of time elapsed.

4. Conclusions

In this paper, we have investigated the impact of the sample size, through an adaptive algorithm, on the solution of the non-stationary newsvendor problem.

This algorithm is well-suited to capture the non-stationarity of the demand in many applications, and ensures that the decision-maker will take immediate action to address change in the underlying demand process, rather than ordering amounts based on historical data that do not reflect customer behavior in the next time period. Future research directions include further fine-tuning of the algorithm, as well as extensions to multiple products, censored demand data and finite time horizon.

References

Bertsimas, Dimitris, and Aurélie Thiele. (2004). A data-driven approach to newsvendor problems. Technical report, Massachusetts Institute of Technology, Cambridge, MA.

Brumelle, Shelby, and Jeffrey McGill. (1993). Airline seat allocation with multiple nested fare classes. *Operations Research,* **41** 127-137.

Ferguson, Thomas. (1996). *A Course in Large Sample Theory,* Chapman & Hall/CRC, Boca Raton, FL.

Gallego, Guillermo, and Ilkyeong Moon. (1993). The distribution-free newsboy problem: Review and extensions. *Journal of the Operational Research Society,* **44** 825-834.

Godfrey, Gregory, and Warren Powell. (2001). An adaptive, distribution-free algorithm for the newsvendor problem with censored demands, with applications to inventory and distribution. *Management Science,* **47** 1101-1112.

Levi, Retsef, Robin Roundy, and David Shmoys. (2006). Provably near-optimal sampling-based policies for stochastic inventory control models. *Proceedings of the 38th annual ACM Symposium on the Theory of Computing (STOC)*, to appear.

Metan, Gokhan, and Aurélie Thiele. (2006). The data-driven newsvendor problem. Technical report, Lehigh University, Bethlehem, PA.

Porteus, Evan. (2002). *Stochastic Inventory Theory*, Stanford University Press, Palo Alto, CA.

Robbins, Herbert, and Sutton Monro (1951). A stochastic approximation method. *Ann. Math. Statis.* **22** 400-407.

Scarf, Herbert. (1958). A min-max solution of an inventory problem, in *Studies in the mathematical theory of inventory and production,* pages 201-209, Stanford University Press, Palo Alto, CA.

van Ryzin, Garrett, and Jeffrey McGill (2000). Revenue management without forecasting or optimization: An adaptive algorithm for determining airline seat protection levels. *Management Science,* **46** 760-775.

A NEIGHBORHOOD SEARCH TECHNIQUE FOR THE FREEZE TAG PROBLEM

Dan Bucantanschi[1], Blaine Hoffmann[2], Kevin R. Hutson[3], and R. Matthew Kretchmar[1]

[1] *Department of Mathematics & Computer Science*
Denison University
Granville, Ohio 43023
kretchmar@denison.edu, bucata_d@denison.edu

[2] *College of Information Sciences and Technology*
Penn State University
University Park, PA 16802
bhoffman@ist.psu.edu

[3] *Department of Mathematics*
Furman University
Greenville, SC 29613
kevin.hutson@furman.edu

Abstract The Freeze Tag Problem arises naturally in the field of swarm robotics. Given n robots at different locations, the problem is to devise a schedule to activate all robots in the minimum amount of time. Activation of robots, other than the initial robot, only occurs if an active robot physically moves to the location of an inactive robot. Several authors have devised heuristic algorithms to build solutions to the Freeze Tag Problem. Here, we investigate an update procedure based on a hill-climbing, local search algorithm to solve the Freeze-Tag Problem.

Keywords: Metaheuristics, swarm robotics, neighborhood search, improvement graph, combinatorial optimization

1. Introduction

Consider the following problem that arises in the field of swarm robotics ([5]). Suppose there are n robots placed in a d-dimensional space. Starting with one initially active robot and the other $n-1$ robots inactive, the goal is to "awaken" the inactive robots so that all n robots are awakened in the fastest time possible. Robot x can awaken robot y only by physically moving to the location of robot y. The cost for robot x to awaken robot y is determined by the straight-line, geometric distance between x's current position and y's position; though not considered here, other variants of this problem constrain robots to travel only on weighted edges of a prescribed graph. Once a robot becomes active, it can assist in awakening the remaining dormant robots. The goal is to compute an optimal awakening schedule, i.e. a schedule that minimizes the time to activate all robots, also known as the *makespan*. Arkin, et. al. [5] dubbed the problem the Freeze-Tag Problem (FTP) for its similarities to a children's game of the same name.

The problem can also be described in the following context. A telecommunications company would like to build a communications network of minimum total cost for the dissemination of information from a single source r to all other network nodes. In wanting to achieve a desired level of service quality, the company constrains itself to build a network with minimum longest path from r to allow for fast, reliable communication links between the source and the customers. Also, the company wants to limit the degree of each node in the network so as to more equally distribute the workload in routing information. A spanning tree network design is a minimum cost alternative for the company because it allows for the desired communcation without redundant network links. Hence the company desires to build a spanning tree network with bounded vertex degrees and minimum longest path from the source. Note that a solution to an instance of the FTP is a spanning tree respresenting the path that robots take with minimum longest path from the initially-awaken robot. At each subsequent awakening, two robots are able to disperse to activate other robots resulting in a degree bound of 3 on the spanning tree solution.

1.1 Related Work

As seen in the second example, the problem of waking a set of sleeping robots in the manner described by the FTP has similarities to various problems arising in broadcasting, routing, scheduling, and network design ([5]). These broadcasting and network design problems share elements of trying to determine how to efficiently disseminate information through a network. Similar problems have arisen in the past to model this data dissemination, such as the minimum broadcast time problem (see [9] for a survey), the multicast problem ([6, 7]), and the minimum gossip time problem ([15]). However, as shown in [5],while the minimum broadcast time problem can be solved in polynomial time in tree networks, the FTP remains intractible even on weighted star graphs.

In fact, much of the prior research on the FTP has focused on proving that it is NP-hard and on designing heuristic algorithms to solve it. Arkin, et. al. ([4]) prove that the FTP is NP-hard on unweighted graphs. These authors show that any "nonlazy" strategy yields an $O(\log n)$ approximation but that an $O(\log n)$ approximation under the same strategy in general metric spaces is difficult to obtain. Sztainberg, et. al. ([17]) prove that a natural greedy heuristic applied to a geometric instance gives an $O((\log n)^{1-\frac{1}{d}})$ approximation in d-dimensions. Their experiments, using several heuristics described in Section 2, show that this greedy approach performs well on a broad range of data sets and yields a small constant-factor approximation.

The FTP is closely related to a more general problem called the *Bounded Degree Minimum Diameter Spanning Tree Problem* (BDST). Introduced in [12], the BDST problem is stated as follows. Given an undirected complete graph $G = (V, E)$ with metric lengths c_{ij} for each $(i, j) \in E$ and bounds $B_v > 0$ on the degree of each $v \in V$, find a minimum-diameter spanning tree T so that for each $v \in V$ the tree-degree of each node v is not greater than B_v. A solution to a Freeze-Tag instance with makespan ρ corresponds to finding a degree-3-bounded spanning tree with longest root-to-leaf path ρ. We should note, however, that the FTP will always have an initial root node with degree 1 which, in general, does not apply to the BDST.

Konemann, et. al. ([11]) provide an $O(\sqrt{\log_B n}) * \Delta$ approximation algorithm for the BDST, where B is the max-degree in the spanning tree and Δ is the minimum diameter of any feasible T. This algorithm provides the best general bound for the FTP as well. This algorithm will also be described in Section 2. Arkin, et. al. ([4]) propose an algorithm that performs better than the algorithm of [11] in some special cases. Namely, for the case where the graph is unweighted, these authors provide an $O(1)$-approximation.

1.2 Outline

The motivation of this paper is to propose a local hill-climbing strategy based on an update graph and search algorithm. We refer to this as the Alternating Path Algorithm for the way it searches a local neighborhood; we will show how this algorithm finds neighboring awakening schedules by alternately adding and removing edges from the current schedule. We compare the performance of the Alternating Path Algorithm on the Freeze-Tag problem against previously published results based on heuristics and combinatorial search strategies.

The paper is outlined as follows. In Section 2, we describe the existing heuristic methods to build approximate solutions for the FTP. We also review two combinatorial search strategies based on genetic algorithms and ant algorithms. We then propose, in Section 3, the Alternating Path Algorithm that employs an improvement graph to take a given awakening schedule and update this solution to a schedule with decreased makespan that still satisfies the degree bounds on each vertex. Finally in Section 4 we present our experimental results.

2. Preliminaries

2.1 Notation

In this section, we describe the basic notation used in the rest of the paper. Some of the graph notation used, such as the definitions of graphs, trees, degrees, cycles, walks, etc., are omitted here, and the reader is referred to the book of Ahuja, Magnanti, and Orlin ([2]). Let $G = (V, E)$ be an undirected network where associated with each edge $e = (i, j)$ is a weight (perhaps metric distance) c_{ij}. Suppose that T is a rooted spanning tree of G with node r specially designated as the root node. Each arc $(i, j) \in E(T)$ denotes a parent-child relationship where i is the parent and j is the child of i. Under this terminology, r is an ancestor to every node in T. Let T_i denote the subtree of T rooted at i.

Associated with each $v \in V(G)$, let $\delta_v(G)$ (also $\delta_v(T)$) denote the degree of v in G (also T) and a number B_v to be a degree bound on v. That is, if T is a spanning tree solution for the FTP, we require $\delta_v(T) \leq 3$, $v \neq r$. Unique to the FTP, is the requirement that the root have degree 1, $B_r = 1$, since the root is the initially active robot and it can only travel to awaken one other robot. From that point forward, there will always be two robots (the already active one, and the newly active one) which can leave a node hence the requirement $\delta_v(T) \leq 3$ representing the path of the incoming robot and the two paths of the outgoing robots. Recall, a path $P = \{v_0, e_0, v_1, e_1, \ldots, v_k - 1, e_{k-1}, v_k\}$ in G is a sequence of nodes and arcs with $e_i = (v_i, v_{i+1}) \in E$ so that no node is repeated. Given a path P, let $dist(P) = \sum_{i=0}^{k-1} c_{v_i, v_{i+1}}$.

2.2 Heuristic Solutions to the FTP

We review some of the existing approaches for building a heuristic solution for the FTP and BDST. Since minimum spanning trees can be built by a greedy approach ([2]), it makes sense to apply this approach to the FTP. Simply stated, under a greedy awakening strategy, once a robot is awakened, it locates the nearest asleep robot to it and attempts to wake it. Sztainberg et. al. [17] explain though that any heuristic that attempts to build a solution from scratch in a greedy fashion must specify how conflicts among robots are resolved, since more that one robot might desire to wake the same neighboring robot. One way to avoid this conflict is to allow robots to claim their next target, and once an inactive robot is claimed by an active robot, it cannot be claimed by another. This greedy approach will be called Greedy Fixed (GF). Alternatively, one could allow claims to be refreshed as needed. Using a greedy approach, a newly active robot could renegotiate the claim of another robot if the newly awakened robot is closer to the claimed inactive robot. This approach combined with an offline delayed target choice to avoid physical oscillations of robots will be refered to as Greedy Dynamic (GD). Experimental results ([17]) show that the greedy dynamic outperforms other methods over a variety of data sets.

Konemann et. al. [11] design an algorithm based on clustering the vertices of the graph. Their idea is to partition the nodes into low-diameter components. They then form a balanced tree which spans the clusters and has a small number of long edges. This ensures that the components are connected by a low-diameter tree. Finally, for each component, a low-diameter spanning tree is found with max-degree B. This divide and conquer approach improves upon the theoretical runtime of the approaches in [17].

2.3 Metaheuristic Solutions to the FTP

A second approach for generating good solutions for the FTP is to apply metaheuristic algorithms to FTP instances. Here, we investigate two such metaheuristic approaches: genetic algorithms and ant colony algorithms. We briefly describe each below, but the interested reader is referred to [1] for complete details on the genetic algorithm and [8] for information on ant colony optimization. Genetic algorithms leverage the principles of Darwinian evolution to stochastically search a complex solution space. These algorithms begin with an initial population of awakening schedules. Each schedule is evaluated by computing its makespan. Based on the makespan, solutions are probabilistically removed from the population (there is a higher probability of retaining schedules with better makespans). Some of the remaining awakening schedules are copied into the new population; others are combined (through a cross-over operator) to form new awakening schedules. Finally, mutations alter some of the solutions in the new population. As the cycle repeats with each subsequent generation, the overall fitness of the population increases and very good solutions are increasingly more likely to be discovered.

Specifically to the FTP, we must define appropriate cross-over and mutation operators for our genetic algorithms. For the cross-over operator, we employ a variant of genetic algorithm cross-over operators used in similar problems [10, 14]. Suppose two parent solutions, T and $\overline{T}$, are to be combined in cross-over to create a child solution. The child's root node is selected to be the root node in the parent with the smaller makespan. The child also contains all edges common to both parents. This results in a graph with k connected components, $T^1, T^2, \ldots, T^k$. For each i, if $|T^i| = 1$, the component is connected to other components by adding in the smaller of the two parent's edges used to connect this node. Lastly, if any robots were not connected to the awakening schedule in the prior step the algorithm searches top-down for the first edge to connect a node to the forest. This is equivalent to trying to place robots in the awakening schedule earlier rather than later. The offspring now replaces the parent with the larger makespan in the next generation of solutions while the parent with the smaller makespan is retained. The mutation operator randomly swaps edges between nodes in an awakening schedule.

Ant algorithms derive their concept from the ways ants search for food. Real-life ants use pheromone trails to guide themselves back and forth from food sources. Ants are attracted to pheromone trials with high levels of pheromone. Those trails which find the shortest path to food sources get traversed faster and therefore have their phermone levels reinforced more often.

For the Freeze-Tag problem, each of m ants are sent out to form an awakening schedule. Each ant will leave a pheromone trial in two places: on the edges of the graph to reflect the chosen awakening schedule and on the nodes of the graph to reflect whether both or only one robot left the node to awaken other robots. Each ant will choose the next robot to be awakened at node j from node i, at iteration t with some probability depending on the following:

1 Which nodes ant k has left to visit. Each ant k keeps track of a set J_i^k of nodes that still need to be processed.

2 The inverse distance d_{ij} from node i to node j called the visibility and denoted $\eta_{ij} = \frac{1}{d_{ij}}$.

3 The amount of virtual pheromone at iteration t representing the learned desirability of choosing node j when transitioning from node i.

4 The amount of virtual pheromone representing the learned desirability to choose to send either one or two robots from node i to awaken other robots.

3. A Neighborhood Search for the FTP

In this section, we introduce an improvement graph structure, similar to [3] and a search method to indicate attractive edge exchanges for solutions of the FTP. Given a solution T to the FTP, an improved solution $\overline{T}$ constitutes a degree-3 bounded tree whose maximum root-to-leaf path is smaller than that of T. More concretely, given any feasible solution $T = (V, E')$ to the FTP problem, with $E' \subseteq E$ and $|E'| = |V| - 1$, let $N^{(k)}(T)$ be the set of feasible trees $\overline{T}$ which differ from T in exactly k edges. The sequence $N^{(1)}(T), N^{(2)}(T), \ldots, N^{(k_{max})}(T)$, $k_{max} \leq |V| - 1$, defines a so-called *neighborhood structure* ([13]) relative to T for the FTP problem. We seek to explore this neighborhood structure for a tree $\overline{T}$ whose makespan is less than T.

Many authors have introduced techniques to perform large-scale neighborhood searches in graphs, see for instance [3, 13, 16]. These methods have been applied to similar problems such as the degree-constrained MST problem [16] and the capacitated MST problem [3]. In the latter, the authors define a corresponding graph, called an improvement graph, that is used to indicate profitable exchanges of vertices and subtrees among multiple subtrees in a current solution to produce an improved solution. Unlike [13] which randomly generates trees in $N^{(i)}(T)$, using this improvement graph gives a deterministic approach to finding improved solutions in the neighborhood structure.

3.1 The Search Method

Our goal is to define an improvement graph and a search technique that allows us to find a tree $\overline{T} \in N^{(i)}(T)$, $i = 1, \ldots, k_{max}$, such that the maximum root-to-leaf path in $\overline{T}$ is less than that of the current solution T. To this end, let $v \in T$ with degree less than B_v be called a *candidate* vertex. Note for the Freeze Tag problem $B_v = 3$, $\forall v \in V$, and a vertex that has 0 or 1 children is a candidate. Let P_j^l be the path and $d^l[j]$ be the distance from vertex j in the tree T to the farthest leaf from j in T_j. Let $d^r[i]$ be the distance along edges in T from the root vertex r to the vertex i. Let $\rho(T)$ denote the length of the longest path, $P_l(T)$, in T from r. We wish to find a tree T' with $\rho(T') \leq \rho(T)$.

We now define the specifics of this improvement graph. Let $G^1(T)$ be a directed graph with vertex set the same as T. Let (i, j) be a directed edge in $G^1(T)$ if either of the following conditions hold. If i is the parent of j in T, then (i, j) is a directed edge in $G^1(T)$ called a *tree edge.* If $(i, j) \notin T$, $j \neq r$, then (i, j) is a directed edge in $G^1(T)$ if j is not a descendant of i in T and a distance condition is satisfied; we call this type of edge a *nontree edge.* This distance condition will be discussed later. The case of $j = r$ is more complicated and will not be considered here.

We wish to show that an exchange of edges/subtrees in T corresponds to traversing an alternating path $AP = \{v_0, e_1, v_1, e_2, \dots, v_{k-1}, e_k, v_k\}$ between tree edges ($E^T = \{e_1, e_3, \dots, e_{k-1}\}$) in $G^1(T)$ and nontree edges ($E^N = \{e_2, e_4, \dots, e_k\}$) in $G^1(T)$ beginning at a vertex v_0 in $P_l(T)$ and ending at a candidate vertex. Note, this ending candidate vertex could be v_0 creating an alternating cycle. If such an alternating path exists, in forming $\overline{T}$ from T we delete those traversed tree edges in $G^1(T)$ from T and add edges (j, i) to $\overline{T}$ whenever edges $(i, j) \in E^N$ are traversed. Define $\overline{E}^N$ to be such that $(j, i) \in \overline{E}^N$ whenever $(i, j) \in E^N$. We claim that $\overline{T} = T - E^T + \overline{E}^N$ is a spanning tree, rooted at r, such that for all $v \in \overline{T}$, $\delta_v \leq B_v$. To illustrate this, consider the following operations for exchanging edges.

EXAMPLE 1 ***Child/Subtree Promote:***

Let i be a non-root vertex on the longest root-to-leaf path such that $\delta_i(T) < B_i$. Let T_i be a subtree of T, and let k, with parent $j \neq i$ be a grandchild of i in T. Further, let $d^l[k] + c_{ik} + d^r[i] < \rho(T)$. Note, if the Triangle Inequality is satified by the edge weights, c_{ik} is less than the tree-path distance between i and k. In the improvement graph, the link (k, i) would be a nontree edge in $G^1(T)$. Hence the alternating path $AP = \{j, (j, k), k, (k, i), i\}$ exists in $G^1(T)$ and lowers the longest root to leaf path since $d^l[k] + c_{ik} + d^r[i] < \rho(T)$,.

EXAMPLE 2 ***Child/Subtree Exchange:***

In this operation, two vertices b and c (non-ancestors) each swap one of their descendant subtrees. This operation is shown in Figure 1 along with alternating cycle $(b - e - c - f - b)$ in the improvement graph. This operation can be extended easily to multiple edge swaps.

We now wish to show by exchanging edges in this fashion that $\overline{T} = T - E^T + \overline{E}^N$ is a rooted spanning tree that satisfies all degree constraints.

LEMMA 3 $\overline{T}$ is a spanning tree rooted at r.

Proof. Let $AP = \{v_0, e_1, v_1, e_2, \dots, v_{k-1}, e_k, v_{k+1}\}$ be an alternating path between tree edges ($E^T = \{e_1, e_3, \dots, e_{k-1}\}$) and nontree edges

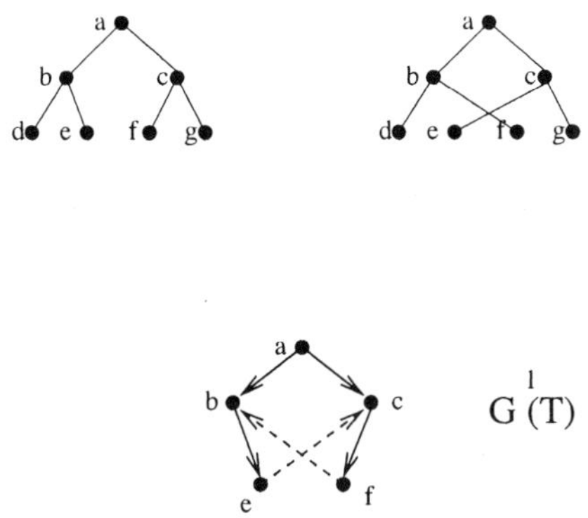

Figure 1. Illustration of Child Exchange and Corresponding Improvement Subgraph

($E^N = \{e_2, e_4, \ldots, e_k\}$) beginning at a vertex v_0 in $P_l(T)$ and ending at a candidate vertex. Since $(j, r) \notin AP$ by construction, the root node cannot change. Thus, we need only to show that $\overline{T}$ is a connected graph with $n-1$ edges. Let $T^1 = T - \{e_1\} + \{e_2\}$, $T^2 = T_1 - \{e_3\} + \{e_4\}$, $\ldots$, $T^{\frac{k}{2}} = \overline{T} = T^{\frac{k}{2}-1} - \{e_{k-1}\} + \{e_k\}$. We wish to show each T^i, $1 < i < \frac{k}{2}$ is a spanning tree rooted at r. We proceed by induction.

Consider T^1. When $e_1 = (i, j)$ is removed from T, T is disconnected into two disjoint sets of connected vertices, $V(T_j)$ and $V(T - T_j)$. Note that since edge $e_2 = (k, j) \in \overline{E}^N$, (j, k) is a non-tree edge in $G^1(T)$, and thus $k \notin V(T_j)$. Thus e_2 connects $V(T_j)$ to $V(T - T_j)$, and hence T^1 is a connected graph. Further, $|E(T)| = |E(T^1)| = n - 1$. Thus T^1 is a spanning tree rooted at r.

Assume T^i, $1 < i < \frac{k}{2}$ is a spanning tree rooted at r. Consider $T^{i+1} = T^i - e_{2i-1} + e_{2i}$. As before, when e_{2i-1} is removed, T^i is disconnected into two disjoint sets of connected vertices which are reconnected by $e_{2i} = (s, t)$ by virtue of s not being a descendant of t in T. Further, since T^{i+1} is formed from T^i by removing one edge and adding one edge back, $|E(T^{i+1})| = n - 1$. Thus T^{i+1} is a spanning tree rooted at r. We conclude $\overline{T}$ is a spanning tree rooted at r. □

LEMMA 4 $\overline{T}$ satisfies $\delta_v \leq B_v$ for all $v \in V(\overline{T})$.

Proof. Note that if $v \notin AP$, then $\delta_v(\overline{T}) = \delta_v(T) \leq B_v$. Also, if v is an interior vertex of AP, it gains a new child and loses an old child and thus $\delta_v(\overline{T}) = \delta_v(T) \leq B_v$. Hence, we need only to consider the end vertices of the path AP: v_0 and v_{k+1}. The vertex v_0 loses a child (v_1) so $\delta_{v_0}(\overline{T}) < \delta_{v_0}(T) \leq B_{v_0}$. Further, by construction of AP, in T, $\delta_{v_{k+1}}(T) < B_{v_{k+1}}$. Thus, since v_{k+1} gains only one child, v_k, $\delta_{v_{k+1}}(\overline{T}) = \delta_{v_{k+1}}(T) + 1 \leq B_{v_{k+1}}$, and the result is shown. □

Given that the improvement graph can be used to generate alternating paths ending at a candidate vertex, one option for generating an improved solution is to successively find such paths and test whether such a path produces an edge exchange that would result in $\rho(\overline{T}) < \rho(T)$. Under this method, the only criteria for a nontree edge $e = (i, j)$ to be included in $G^1(T)$ would be that j is not a descendant of i. Another option would be to establish a distance criteria for nontree edge inclusion in $G^1(T)$ so that the improvement graph might be used to indicate attractive edge exchanges. One obvious criteria would be that if $(i, j) \notin T$, $j \neq r$, then (i, j) is a directed nontree edge in $G^1(T)$ if $d^l[j] + c_{ij} + d^r[i] < \rho(T)$. This criteria, though, is not reliable under all search methods of $G^1(T)$. Consider a nontree edge $(u, v) \in G^1(T)$. If a search method for $G^1(T)$ traverses (u, v) then T_u will be attached to v in T'. However, this attachment could change the distance labels $d^l[j]$ for v and its ancestors and $d^r[i]$ for u and its descendants making their distance labels unreliable. One alternative to combat this is to change the distance to the farthest leaf or distance to the root for nodes as we traverse paths in $G^1(T)$. A second alternative is to attempt to search edges $(i, j) \in G^1(T)$ where $d^l[j] + c_{ij} + d^r[i] < \rho(T)$ is guaranteed to hold. To accomplish this, we restrict movement in $G^1(T)$ to be between disjoint subtrees. This restriction is enforced in [3] under a natural condition of a capacited minimum spanning tree. Here, more care is needed. Let P be an alternating path (or cycle) in $G^1(T)$, we say $P = \{v_0, e_0, v_1, e_1, \ldots, v_k, e_k, v_{k+1}\}$ between tree edges $E^T = \{e_0, e_2, \ldots, e_{k-1}\}$ and nontree edges $E^N = \{e_1, e_3, \ldots, e_k\}$ in $G^1(T)$. Then P is a *reliable path* if for each $e_i = (v_i, v_{i+1}) \in E^N$, every v_j, $j \geq i + 1$, is neither an ancestor of v_{i+1} nor a descendant of v_i. This extra condition ensures the reliability of the distance labels. Given a reliable path (cycle) in $G^1(T)$, we can show $\overline{T}$ is a degree-bounded rooted spanning tree with no greater maximum root-to-leaf path than T.

The improvement graph takes $O(n^2)$ time to construct since each vertex pair must be checked to determine whether the conditions for edge inclusion have been satisfied. To check whether a vertex is an ancestor or descendant of a previously visited vertex, a $0-1$ ancestor and descendent array is employed. This makes this check excutable in $O(1)$ time. The complexity of finding such an alternating path or cycle of length less than $2 \cdot k_{max}$ (indicating $\overline{T}$ has k_{max} edges different from T) is $O(k_{max} \cdot n)$ per node used as v_0. Searching for tree/nontree edges from each node involves just a scan of the edges emanating from the node taking $O(n)$ time per node. To produce an alternating path of length $2k_{max}$ then is accomplished in breadth-first fashion in $O(k_{max} \cdot n)$ time. We limit ourselves to choosing the initial vertex on AP to lie on the path defining $\rho(T)$. So our search takes $O(k_{max} \cdot n^2)$ time. In implementation, we limit $k_{max} \leq 4$.

4. Experimental Results

In this section we illustrate the generation of these alternating paths in the improvement graph. This algorithm will be referred to as the Alternating Path Algorithm (AP). We then discuss implementation details of the algorithm such as search depth (previously referred to as k_{max}) and iterative improvement. We compare the results of AP against other benchmarks noted in the previous work discussion.

4.1 Illustrating the Alternating Path Algorithm

For this discussion, we use the Eil51 problem from the TSPLIB [18] which contains 51 robots dispersed in 2D space with $\{x, y\}$ coordinates in the range of $(0, 0)$ to $(70, 70)$. The location of these robots is shown in Figure 2. The labels of these robots is arbitrary (set by the Eil51.tsp file) but are useful for referring to specific robots.

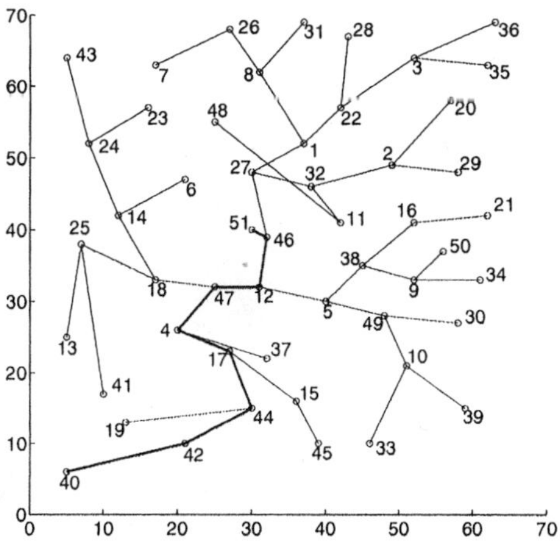

Figure 2. Greedy Dynamic's Solution for Eil51.tsp

Since the Alternating Path Algorithm improves upon an existing solution to the FTP, we need an awakening schedule to use as a "seed" value. We use Greedy Dynamic (see Heuristic solutions discussed earlier) as an algorithm that provides a starting point with a good, but not nearly optimal awakening schedule. The structure of Greedy Dynamic's solution is shown in Figure 2. This awakening schedule has a makespan of $\rho(T) = 66.07$ seconds with the root node in the middle (rootID = 51); The rootID of 51 was chosen because this produces the minimal makespan over all possible root nodes using the Greedy Dynamic algorithm (though this may not be the best choice for a root for the true optimal awakening schedule). The longest root-to-leaf path is indicated by the shaded edges in Figure 2.

Mechanically, the Alternating Path Algorithm tries to find reliable alternating paths in $G^1(T)$ along the longest root-to-leaf path. It first starts from robot 46, and then continuing successively along the longest root-to-leaf path, it searches for reliable alternating paths of depth k, $1 \leq k \leq k_{max}$, from robot 12, robot 47, etc. The algorithm exhaustively considers all such reliable alternating paths of a certain depth from each of these nodes. Of all these new paths found (and corresponding new trees generated), the best is recorded and is used as the solution returned by AP.

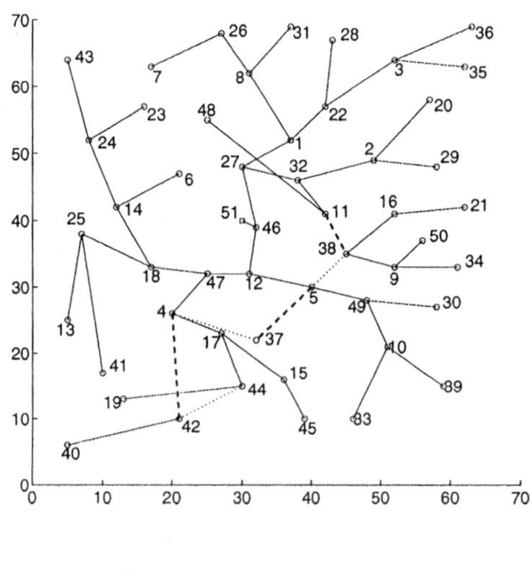

Figure 3. Illustration of Alternating Path Algorithm

For the specific case in Figure 2, the best alternating path structure was found using the connection between robots 44 and 42. In Figure 3 we see the newly created awakening schedule. The edges which have been added are shown with heavy dashed lines. The edges which were removed are shown in light dotted lines. The alternating path discovered here consists of $\{ (44, 42), (42, 4), (4, 37), (37, 5), (5, 38), (38, 11)\}$ where the odd edges are the ones removed and the even edges are the ones added. The makespan of this new spanning tree is $\rho(T') = 56.80$ seconds (compared with 66.07 seconds of the seed tree). Note that the longest path has changed, it now starts at the root and progresses toward the upper left corner: $(51, 46, 12, 47, 18, 14, 24, 43)$.

4.2 Depth Control, Local Search Topology, and Iteration

The *Iterative Alternating Path* algorithm (IAP) simply repeats application of the AP algorithm on each successive solution T' generated until no further improvements can be made. As shown in Figure 3, the 1-step AP algorithm produced a makespan of 56.8 seconds with a new longest path in the upper-left corner. Running AP again on this new tree is likely to generate an even better solution. For this specific instance, we find that an additional 14 iterations are possible with incremental improvements at each step using IAP on Eil51. The final awakening schedule (shown in Figure 4) has a makespan of 51.57.

As discussed previously, $N^{(k)}(T)$ is the neighborhood of all trees differing from T by k edges. This actually infers a pair of edges (one removed and one added) traversed by AP. Thus, the example awakening schedule illustrated in Figure 3 belongs to $N^{(3)}(T)$. By controlling the k_{max} parameter, we can limit the number of edge exchanges in our search process.

The parameter, k_{max}, determines the number of edge exchanges we may consider to construct the neighborhood of a solution T. Naturally, increasing the parameter k_{max} expands the neighborhood thereby allowing us to consider a greater number of potential next solutions and possibly discovering a better local makespan that would not have been found in a single step with a lower k_{max}. If the local search topology is a smooth basin, without many local minima, then larger k_{max} is generally better. However, if the topology is more varied, then larger steps may take us out of of local basin where a very good solution might lie; or alternately, allow us to skip into a better basin with a better local minima.

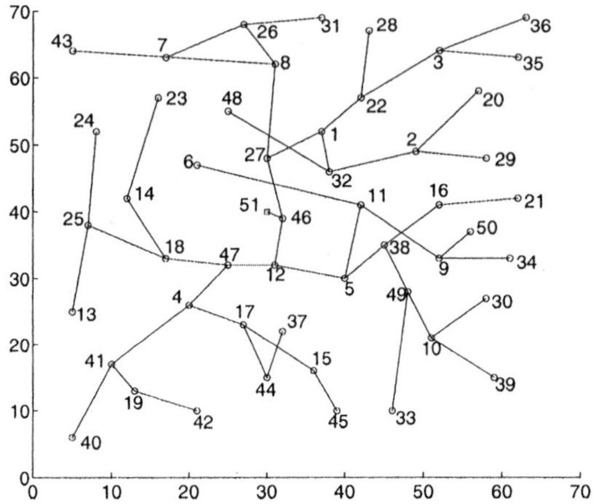

Figure 4. Iterative Alternating Path Algorithm

An intriguing result allows us to hypothesize that the topology is probably more varied than smooth. On the Eil51 problem, we ran Iterative Alternating Path with a depth of $k_{max} = 3$ and achieved a best solution of 51.57 after 15 iterations. Increasing $k_{max} = 4$ drops the number of iterations down to just seven. However, the locally optimal solution found has a makespan of only 53.69. The larger neighborhood structure of this latter run allowed us to accept an intermediate solution that was not available with the smaller neighborhood, but by doing so, it stepped out of a basin that contained ultimately a better solution (51.57).

4.3 Results

AP is clearly a local hill-climbing method; it finds solutions within a neighborhood of its current best solution and then chooses the most optimal among these. Furthermore, it is limited to only improving steps and thus will never escape a local minima in its neighborhood of search. The more robust searching algorithms (ants and GAs) are more adapt at escaping local minima and exploring a larger portion of the solution space.

The IAP algorithm is also computationally intensive. Running on a fast workstation, a 200 robot problem with depth=4 requires more than 24 hours of computation time. Increasing the problem size even modestly (say `lin318.tsp` with 318 robots) prohibits full computation; we are limited to computing only a subset of initial seeds.

This suggests a more proper role for AP/IAP. Instead of using AP as a basic search algorithm, it is better to use a more exploratory algorithm first. Then AP can be used as a final or intermediate step operating on the best (or best few) solution found with the first search algorithm. Thus IAP acts more like a last-step improvement procedure to further realize gains on an already good solution.

The following table summarizes the best solutions found using each method on seven different FTPs using the IAP and comparing it with some of the heuristic and metaheuristic algorithms discussed in the Previous Work section. These seven problems were selected from the TSPLIB (Traveling Salesperson Problem Library) [18]. In the table, Alternating Path indicates the best solution found by AP using the best 10 solutions found by Greedy Dynamic as initial seeds. We also employed a primitive version of AP (with $k_{max} \leq 2$) as an intermediate step to further improve upon the solutions that the ant algorithm had discovered; thus the results for ant algorithm are a bit better than would normally be accomplished with the algorithm alone. The Genetic Algorithm search also performs well. We can see in general that IAP and the metaheuristic algorithms find solutions which are significantly better than Greedy Dyanmic, and using the IAP as an intermediate step in the ant algorithm affords significant improvements over the initial seed solutions. This suggests that the IAP algorithm is useful in creating improved solutions for the FTP used either with Greedy Dynamic or in conjunction with a metaheuristic approach such as ant algorithms.

Table 1. Best Solution found using various algoithms.

DataSet → **Algorithm** ↓	eil51	eil76	kroA100	d198	lin318	att532	rat783
Greedy Fixed	75.03	68.7	3857	3087	4612	9720	463.3
Greedy Dynamic	61.63	56.64	2515	2433	2806	5096	340.8
Center of Mass	55.92	59.65	2591	2446	2944	5367	362.6
Genetic Algorithm	54.79	51.72	2393	2445	2759	5064	475.7
Ant Algorithm	49.20	50.28	2396	2380	2700	5073	336.0
Alternating Path	50.56	52.77	2442	2420	2717	4989	335.4

References

[1] Aarts, E., and J. Lenstra. (1997). *Local Search in Combinatorial Optimization*, John Wiley and Sons, Ltd.

[2] Ahuja, R., T. Magnanti, and J. Orlin. (1993). *Network Flows: Theory, Algorithms, and Applications*. Prentice Hall, Englewood Cliffs, NJ.

[3] Ahuja, R., J. Orlin, and D. Sharma. (1997) Multi-exchange neighborhood search algorithms for the capacitated minimum spanning tree problem. *Mathematical Programming*, 91:71-97.

[4] Arkin, E., M. Bender, G. Dongdong, S. He, and J. Mitchell. (2003). Improved Approximation Algorithms for the Freeze-Tag Problem. In 15th *ACM Sym. of Par. in Alg. and Arch. (SPAA'03)*, 295-303.

[5] Arkin, E., M. Bender, S. Fekete, J. Mitchell, and M. Skutella. (2002). The Freeze-Tag Problem: How to wake up a swarm of robots. In *Proc. 13th ACM-SIAM Sympos. Discrete Algorithms*, 568-577.

[6] Banerjee, S., C. Kommareddy, K. Kar, B. Bhattacharjee, and S. Khuller. (2003) Construction of an Efficient Overlay Multicast Infrastructure for Real-time Applications. In *INFOCOM 2003.* 22nd *Annual Joint Conf. of the IEEE Comp. and Comm. Soc.*, 2:1521-1531.

[7] Bar-Noy, A., S. Guha, J. Naor, and B. Schieber. (1998). Multicasting in heterogeneous networks. In *Proc. 30th ACM Sympos. Theory Comput.*, 448-453.

[8] Bonabeau, E., M. Dorigo, and G. Theraulaz. (1999). *Swarm Intelligence: From Natuaral to Artificial Systems.* Oxford University Press.

[9] Hedetniemi, S. M., S. T. Hedetniemi, and A. Liestman. (1988). A survey of gossiping and broadcasting in communication networks. *Networks*, 18:319-349.

[10] Julstrom, B. and G. Raidl. (2003). A Permutation-coded evolutionary algorithm for the bounded-diameter minimum spanning tree problem. In *2003 Genetic and Evolutionary Computation Conference's Workshop Proceedings, Workshop on Analysis and Design of Representations*, 2-7.

[11] Konemann, J. and R. Ravi. (2002). A matter of degree: Improved approximation algorithms for the degree-bounded minimum spanning trees. *SIAM J. Comput.*, 31(6):1783-1793.

[12] Konemann, J. and R. Ravi. (2003). Primal-dual algorithms meets local search: Approximating MST's with nonuniform degree bounds. In *Proc. of the* 35th *ACM Symposium on Theory of Computing*, 389-395.

[13] Mladenovic, N. and P. Hansen. (1997). Variable neighborhood search. *Comput. Oper. Res.*, 24:1097-1100.

[14] Raidl, G. (2000). An efficient evolutionary algorithm for the degree-constrained minimum spanning tree problem. In *Proc. of the 2000 Congress on Evolutionary Computation CEC00*, 1:104-111.

[15] Ravi, R. (1994). Rapid rumor ramification: Approximating the minimum broadcast time. In *Proc.* 35th *Sympos. Found. Computer Science*, 202-213.

[16] Ribeiro, C. and M. Souza. (2002). Variable neighborhood search for the degree-constrained minimum spanning tree problem. *Discrete Applied Mathematics*, 118:43-54.

[17] Sztainberg, M., E. Arkin, M. Bender, and J. Mitchell. (2002). Analysis of heuristics for the Freeze-Tag Problem. In *Proc. Scandianvian Workshop on Algorithms*, 2368:270-279.

[18] TSPLIB. http://www.iwr.uni-heidelberg.de/groups/comopt/software/TSPLIB95/

THE COLORFUL TRAVELING SALESMAN PROBLEM

Yupei Xiong[1], Bruce Golden[2], and Edward Wasil[3]
[1]Goldman, Sachs & Co., 85 Broad Street, New York, NY 10004; [2]R.H. Smith School of Business, University of Maryland, College Park, MD 20742; [3]Kogod School of Business, American University, Washington, DC 20016

Abstract: Given a connected, undirected graph *G* whose edges are labeled (or colored), the colorful traveling salesman problem (CTSP) seeks a Hamiltonian tour of *G* with the minimum number of distinct labels (or colors). We prove that the CTSP is NP-complete and we present a heuristic algorithm and a genetic algorithm to solve the problem.

Key words: Genetic algorithm; NP-complete; Hamiltonian tour

1. INTRODUCTION

In the colorful traveling salesman problem (CTSP), we are given an undirected complete graph with labeled edges as input. Each edge has a single label and different edges can have the same label. We can think of each label as a unique color. The goal is to find a Hamiltonian tour with the minimum number of labels. Consider the following hypothetical scenario. An individual wants to visit *n* cities without repetition and return to the city of origin. Suppose that all pairs of cities are directly connected by railroad or bus and that there are *l* transport companies. Each company controls a subset of the railroad and bus lines (edges) connecting the cities and each company charges the same flat monthly fee for using its lines. We can draw the lines owned by Company 1 in red, the lines owned by Company 2 in blue, and so on. The objective is to construct a Hamiltonian tour that uses the smallest number of colors.

Chang and Leu (1997) introduced the Minimum Labeling Spanning Tree (MLST) problem, which they proved to be NP-hard. They also provided a fast and efficient heuristic called MVCA. Since then, the MLST problem has been studied by numerous researchers. In particular, Krumke and Wirth (1998) proposed a modification to MVCA and proved that

MVCA can yield a solution no greater than 2 $ln\ n$ +1 times optimal, where n is the number of nodes. Wan, Chen, and Xu (2002) obtained an improved bound 1 + $ln(n\text{-}1)$. Xiong, Golden, and Wasil (2005a) further improved the performance guarantee to H_b for any graph with label frequency bounded by b (i.e., no label occurs more than b times in G), where $H_b = \sum_{i=1}^{b} \frac{1}{i}$ is the bth harmonic number. Next, they presented a worst-case family of graphs for which the MVCA solution is exactly H_b times the optimal solution. In a subsequent paper, Xiong, Golden, and Wasil (2005b) provided a one-parameter genetic algorithm to solve the MLST problem. In general, their genetic algorithm outperformed MVCA.

In this paper, we prove that the CTSP is NP-complete. Next, we introduce the Maximum Path Extension Algorithm (MPEA) to solve the CTSP. MPEA is a greedy algorithm. We also introduce a genetic algorithm (GA) to solve the CTSP. This GA combines MPEA with the genetic algorithm presented by Xiong, Golden, and Wasil (2005b). It obtains better computational results than MPEA alone.

We point out that a wide variety of network design problems have been addressed using genetic algorithms. For example, see Chou, Premkumar, and Chu (2001), Palmer and Kershenbaum (1995), and Golden, Raghavan, and Stanojevic (2005).

2. THE CTSP IS NP-COMPLETE

Problem: Given a complete graph $G = (V, E, L)$, where V is the set of nodes, E is the set of edges, L is the set of labels, and each edge in E is assigned a label in L, find a Hamiltonian tour that contains the smallest number of distinct labels.

Theorem 1. *The CTSP is NP-complete.*

Proof. First, we show that the CTSP belongs to NP. Given an instance of the problem, we use as a certificate the sequence of n nodes in the tour. The verification algorithm confirms that this sequence contains each node exactly once, sums up the number of labels, and checks whether the sum is at most k. This process can certainly be done in polynomial time.

To prove that the CTSP is NP-complete, we show that HAM-CYCLE can be transformed to the CTSP in polynomial time. HAM-CYCLE is the Hamiltonian-cycle (or Hamiltonian tour) problem which is known to be NP-complete. Let $G = (V, E)$ be an instance of HAM-CYCLE. We construct an instance of CTSP as follows. We form the complete graph $G' = (V, E')$, where $E' = \{(i, j) : i, j\ ?\ V \text{ and } i\ ?\ j\}$, and define the labels as follows.

All the edges in E have the same label c. Each edge in $E' - E$ has a unique label. The instance of CTSP is easily formed in polynomial time.

We now show that graph G has a Hamiltonian tour if and only if graph G' has a tour with only one label. Suppose that graph G has a Hamiltonian tour h. Each edge in h belongs to E and thus has label c in G'. Thus, h is a tour in G' with only one label c. Conversely, suppose that graph G' has a tour h' with only one label. Since each edge in $E' - E$ has a unique label, if h' has at least 2 edges, then all the edges in h' must be in E, and, thus, h' is a Hamiltonian tour in graph G. If h' has only one edge, then G has only 2 nodes. So, we must have $E = E'$ and h' is also a Hamiltonian tour in graph G. Therefore, the CTSP is NP-complete.

An example of the CTSP is presented in Figure 1. In this example, $G = (V, E, L)$ is a complete graph where $V = \{1, 2, 3, 4, 5, 6\}$ and $L = \{a, b, c, d, e, f\}$. Two tours are displayed. Tour h contains three distinct labels and tour g contains two distinct labels. So, tour g is a better solution.

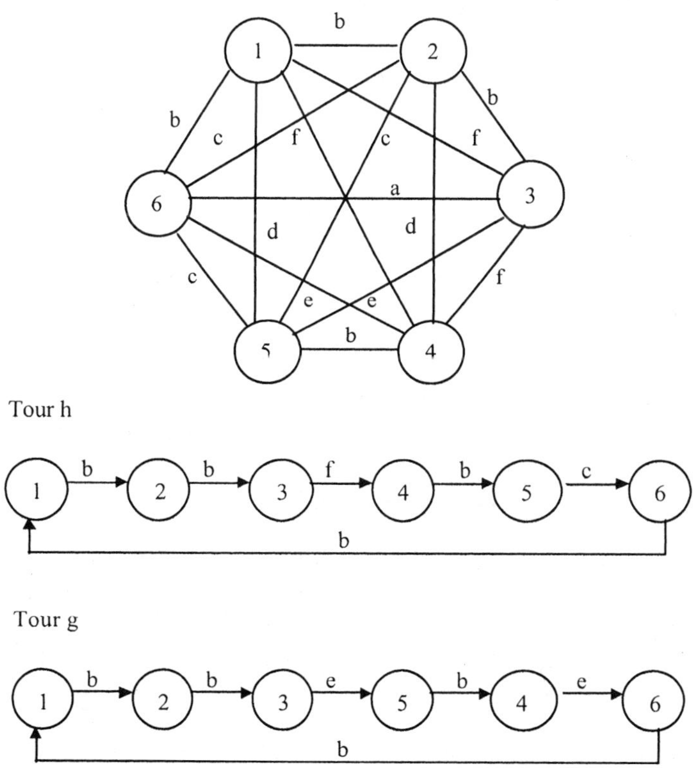

Fig.1. An illustration of the CTSP. Tour h contains the three distinct labels b, c, and f. Tour g contains the two distinct labels b and e. So, tour g is better.

3. MAXIMUM PATH EXTENSION ALGORITHM

In this section, we introduce the Maximum Path Extension Algorithm to solve the CTSP. The basic idea of MPEA is to visit as many nodes as possible while keeping the number of labels the same as in the current partial tour.

3.1 How to extend a partial tour?

In a complete labeled graph $G = (V, E, L)$, suppose we have a partial tour $h : v_1$? v_2 ? . . . ? v_k. Let C_h be the set of labels in the partial tour h. We want to extend the partial tour h to h' by one more node v_{k+1} such that $C_{h'} = C_h$, where $C_{h'}$ is the set of labels in h'. In the trivial case (Case 0), we can find an unvisited node v_{k+1}, such that the label of the edge (v_k, v_{k+1}) or the label of the edge (v_1, v_{k+1}) belongs to C_h. Then, we directly insert v_{k+1} after v_k or prior to v_1. The following four nontrivial cases can also make this extension possible.

Case 1: If we can find some unvisited node v_{k+1} and some $v_j \in h$, such that the labels of the edges (v_j, v_{k+1}) and (v_{j+1}, v_{k+1}) belong to C_h, then we can insert the node v_{k+1} into the partial tour h without increasing $|C_h|$.

Case 2: If we can find some unvisited node v_{k+1} and some $v_j \in h$, such that the labels of the edges (v_{k+1}, v_{j+1}) and (v_j, v_k) belong to C_h, then we can insert v_{k+1} into the partial tour h without increasing $|C_h|$ and v_{k+1} will be one end node of the new tour.

Case 3: If we can find some unvisited node v_{k+1} and some $v_j \in h$, such that the labels of the edges (v_{k+1}, v_j) and (v_1, v_{j+1}) belong to C_h, then we can insert v_{k+1} into the partial tour h without increasing $|C_h|$ and v_{k+1} will be one end node of the new tour.

Case 4: If the label of the edge (v_1, v_k) belongs to C_h, and we can find some unvisited node v_{k+1} and some v_j in h, such that the label of the edge (v_j, v_{k+1}) belongs to C_h, then we can insert v_{k+1} into the partial tour h without increasing $|C_h|$. v_{k+1} and v_{j+1} will be the two end nodes of the new tour.

If any unvisited node cannot satisfy any of the above trivial or nontrivial cases, then we extend the partial tour h by inserting a random unvisited node v_{k+1} at the end (after v_k) of the tour h and add the new label of the edge (v_k, v_{k+1}) to C_h. We can also select an unvisited node v_{k+1} with the highest frequency. This makes sense because this selection is likely to provide many opportunities to succeed in further path extensions.

3.2 Maximum Path Extension Algorithm

Suppose the input is a complete labeled graph $G = (V, E, L)$ with $|V| = n$ nodes. We want to output a Hamiltonian tour h. Suppose C is the set of labels in h. A detailed description of MPEA follows.

Step 1: Sort all the labels in G according to their frequencies, from largest to smallest.

Step 2: Randomly select $v_1 \in V$, then find $v_2 \in V$ such that the label c_{12} of the edge (v_1, v_2) has the highest frequency.

Step 3: Let $h = \{v_1, v_2\}$ and $C = \{c_{12}\}$.

Step 4: Add unvisited nodes to h according to the rules in Section 3.1 (from Case 0 to Case 4), until h contains all n nodes.

Step 5: Suppose $h = \{v_1, \ldots, v_n\}$ is an ordered sequence of nodes, and let label c_{ln} denote the label of the edge (v_1, v_n). If c_{ln} is not in C, then add it to C.

Step 6: Output h.

Now we consider the example in Figure 1 and show how MPEA works for this example. In this example, $V = \{1, 2, 3, 4, 5, 6\}$ and $L = \{b, c, f, d, e, a\}$, where L is sorted in decreasing order of label frequency. We begin with node 1 and we select 2 as the second node because the label of the edge (1,2) is b, which has the highest frequency. So $h = \{1, 2\}$ and $C = \{b\}$. By the rules in Section 3.1, h can be extended to $h = \{6, 1, 2, 3\}$ with C unchanged. We select node 4 next, because the label $c_{34} = f$ is the highest frequency label available from node 3. This yields $h = \{6, 1, 2, 3, 4\}$ and $C = \{b, f\}$. Next, we extend h without increasing the number of labels to obtain $h = \{6, 1, 2, 3, 4, 5\}$ since $c_{45} = b$. Now h visits each node in G. Since $c_{56} = c$, we add it to C to obtain $C = \{b, f, c\}$. We now have a Hamiltonian tour h with three labels, as shown in Figure 1.

3.3 Running time analysis

Given a complete labeled graph $G = (V, E, L)$, let $|V| = n$, $|L| = l$, and $|E| = \frac{n(n-1)}{2} = O(n^2)$. We assume that l and n are of the same order of magnitude. In Step 1, we use quicksort and it requires $O(l \ln l)$ running time. Step 2 requires $O(n)$ running time. Step 3 requires constant running time. Step 4 is the main step of MPEA. In each loop, suppose h contains k nodes $v_i, \ldots, v_k$, so there are $n - k$ unvisited nodes. For each unvisited node u, we need to check the label of edges (u, v_i) for each $1 = i = k$ and determine whether we can extend h to u without changing C. So it requires $O((n - k)k)$ running time. If we succeed in extending h to u, then the following insertion operation requires $O(k)$ running time; if we fail to extend h to all unvisited nodes, then we select the unvisited node with the highest frequency and it requires $O(n - k)$ running time. The loop will be repeated at most n times and k goes from 2 to n. Thus, the total running time of Step 4 is $O(n^3)$. Step 5 requires $O(1)$ running time. Therefore, the total running time for MPEA is $O(n^3)$.

4. GENETIC ALGORITHM

In this section, we introduce a genetic algorithm to solve the CTSP. We use MPEA, but we begin with a selective label set C that contains more than one label. The subgraph H induced by C should be connected and span all the nodes of G.

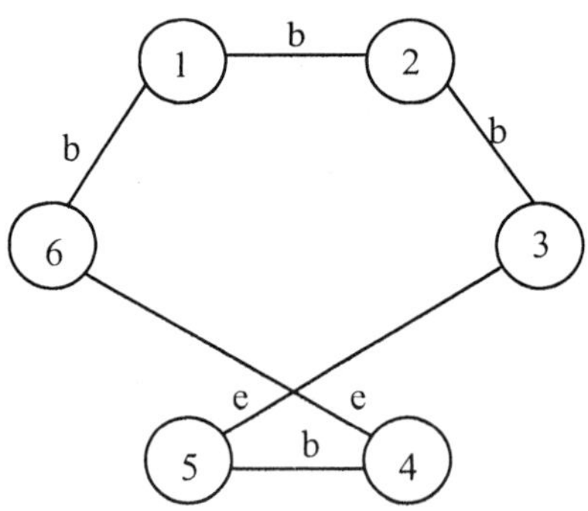

Fig. 2. Subgraph induced by $\{b, e\}$. It contains the Hamiltonian tour 6 ? 1 ? 2 ? 3 ? 5 ? 4 ? 6. This tour has two labels.

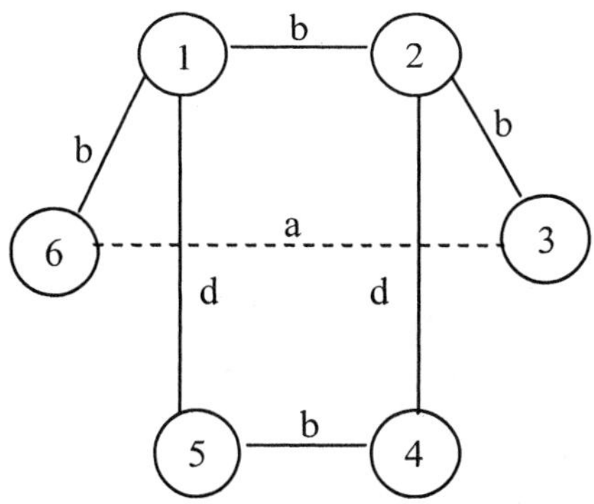

Fig. 3. Subgraph induced by $\{b, d\}$. It does not contain a Hamiltonian tour. We have to add edge (3, 6) to form the Hamiltonian tour 6 ? 1 ? 5 ? 4 ? 2 ? 3 ? 6. This tour contains three labels.

Table 1. Computational results of MPEA and GA

	MPEA	Avg. time (sec)	GA	Avg. time (sec)
$n = 50, l = 25$	2.4	0.1	2.4	0.3
$n = 50, l = 50$	4.5	0.1	4.2	0.4
$n = 50, l = 75$	5.6	0.1	5.7	0.5
$n = 50, l = 100$	6.6	0.1	6.8	0.6
$n = 100, l = 50$	3.5	0.3	3.0	0.9
$n = 100, l = 75$	4.0	0.3	4.1	1.2
$n = 100, l = 100$	5.8	0.2	5.1	1.5
$n = 100, l = 125$	6.3	0.3	6.9	1.7
$n = 100, l = 150$	7.2	0.3	6.9	1.7
$n = 150, l = 75$	3.4	0.7	3.0	5.1
$n = 150, l = 100$	4.5	0.8	4.1	6.2
$n = 150, l = 150$	5.9	0.9	5.5	7.6
$n = 150, l = 200$	7.5	0.9	7.3	8.9
$n = 200, l = 100$	3.8	1.5	3.4	10.1
$n = 200, l = 150$	5.5	1.9	4.9	13.0
$n = 200, l = 200$	6.9	1.9	6.2	14.7
$n = 200, l = 250$	8.2	2.0	7.4	17.2

Given this label set C, extending a partial tour in H is more likely to succeed than if we start with a single label. We are now trying to find a Hamiltonian tour containing approximately $|C|$ labels. So $|C|$ should be as small as possible. We can solve the MLST problem here to obtain C.

Recently, Xiong, Golden, and Wasil (2005b) applied a one-parameter genetic algorithm to solve the MLST problem and we use their genetic algorithm in our procedure.

For the example in Figure 1, if we apply this GA (which begins with the genetic algorithm from Xiong, Golden, and Wasil (2005b) and is followed by MPEA), we might find one of two optimal subgraphs. One is induced by the label set {*b*, *e*} and this subgraph contains a Hamiltonian tour. See Figure 2. So this tour contains two labels, which is the optimal solution. The other is induced by the label set {*b*, *d*}. But this subgraph only contains a Hamiltonian path (6 ? 1 ? 5 ? 4 ? 2 ? 3). After we connect the head and the tail of the path, a new label is created, as shown in Figure 3. So, here we can find a Hamiltonian tour with three labels.

5. COMPUTATIONAL RESULTS

In this section, we give the computational results of MPEA and GA. For each input *n* and *l*, we randomly generate 10 graphs. For each graph, we run MPEA 200 times and find the best result. We run the GA only once and record the best result in the last generation. Finally, we output the average number of labels of the 10 graphs for each input. The computational results are presented in Table 1. In this table, the inputs *n* and *l* are presented in the first column. The average number of labels from MPEA and the associated average running time are given in the second and third columns. In the fourth and fifth columns, the average number of labels from the GA and the associated average running time are presented. There are 17 cases in this computational experiment. The GA outperforms the MPEA in 12 cases, underperforms the MPEA in 4 cases, and ties in one case. The GA is clearly slower than the MPEA. These experiments were run on a Pentium 4 PC with 1.80 GHz and 256 MB RAM.

6. CONCLUSIONS

In this paper, we introduced the colorful traveling salesman problem. We were able to show that the CTSP is NP-complete. Next, we presented a heuristic algorithm (MPEA) and a genetic algorithm (GA) to solve the CTSP. The MPEA is very fast, but the GA yields better results and its running time is still reasonable. This is another nice example of the ability of genetic algorithms to successfully solve difficult combinatorial optimization problems.

7. REFERENCES

Chang, R. S. and Leu, S.-J., 1997, The minimum labeling spanning trees. *Inform. Process. Lett.,* **63**: 277-282.

Chou, H., Premkumar, G., and Chu, C.-H., 2001, Genetic algorithms for communications network design- An empirical study of the factors that influence performance. *IEEE Transactions on Evolutionary Computation*, **5**: 236-249.

Golden, B., Raghavan, S., and Stanojevic, D., 2005, Heuristic search for the generalized minimum spanning tree problem. *INFORMS Journal on Computing*, **17**(3):290-304.

Krumke, S. O., and Wirth, H. C., 1998, On the minimum label spanning tree problem. *Inform. Process. Lett.*, **66**:81-85.

Palmer, C. C. and Kershenbaum, A., 1995, An approach to a problem in network design using genetic algorithms. *Networks*, **26**:151-163.

Wan, Y., Chen, G., and Xu, Y., 2002, A note on the minimum label spanning tree. *Inform. Process. Lett.*, **84**:99-101.

Xiong, Y., Golden, B., and Wasil, E., 2005a, Worst-case behavior of the MVCA heuristic for the minimum labeling spanning tree problem. *Operations Research Lett.,* **33**:77-80.

Xiong, Y., Golden, B., and Wasil, E., 2005b, A one-parameter genetic algorithm for the minimum labeling spanning tree problem. *IEEE Transactions on Evolutionary Computation,* **9**:55-60.

SOLVING THE MULTI-DEPOT LOCATION-ROUTING PROBLEM WITH LAGRANGIAN RELAXATION

Zeynep Özyurt[1] and Deniz Aksen[2]
[1]Industrial Engineering Deparment, Koç University; [2]College of Administrative Sciences and Economics, Koç University, Rumelifeneri yolu 34450 Sariyer, Istanbul, Türkiye

Abstract: Multi-depot Location-Routing Problem (MDLRP) is about finding the optimal number and locations of depots while allocating customers to depots and determining vehicle routes to visit all customers. In this study we propose a nested Lagrangian relaxation-based method for the discrete uncapacitated MDLRP. An outer Lagrangian relaxation embedded in subgradient optimization decomposes the parent problem into two subproblems. The first subproblem is a facility location-like problem. It is solved to optimality with Cplex 9.0. The second one resembles a capacitated and degree constrained minimum spanning forest problem, which is tackled with an augmented Lagrangian relaxation. The solution of the first subproblem reveals a depot location plan. As soon as a new distinct location plan is found in the course of the subgradient iterations, a tabu search algorithm is triggered to solve the multi-depot vehicle routing problem associated with that plan, and a feasible solution to the parent problem is obtained. Its objective value is checked against the current upper bound on the parent problem's true optimal objective value. The performance of the proposed method has been observed on a number of test problems, and the results have been tabulated.

Key words: location routing; Lagrangian relaxation; heuristics; tabu search.

1. INTRODUCTION AND LITERATURE SURVEY

Location-Routing Problem (LRP) involves finding the optimal number and locations of depots while allocating customers to depots and determining vehicle routes to visit all customers. Another problem which establishes depots considering demand and location data of customer nodes is the classical location/allocation problem (LAP). The main difference of the LRP from the LAP is that, once the facilities have been placed, the LRP requires the visitation of demand nodes through tours, whereas the latter assumes straight-line or radial trips between the facilities and respective customers. The LRP considers three main decisions of different levels simultaneously: location of depots - strategic level; allocation of customers to depots - tactical level and the routes to visit these customers - operational level. The interdependence between these decisions has been noticed by researchers long ago. The effect of ignoring routes when locating depots has also been stressed by Salhi and Rand (1989). However, due to the complexity of both location and routing problems, these two have been traditionally solved separately. In the literature there exist heuristic solution methods – for example; Tüzün and Burke (1999), Wu et al. (2002), Albareda-Sambola et al. (2005) – as well as exact methods – for example; Laporte et al. (1988) – proposed for solving the LRP as a whole.

The newest annotated literature review of the LRP and its extensions is currently due to Ahipasaoglu et al. (2004). A complete synthesis and survey of the LRP was accomplished earlier by Min et al. (1998) who propose a classification scheme for LRPs. The authors argue that sequential methods consisting of decomposition for the LRP have their limitation. They recommend solving the subproblems of the LRP concurrently in order to be able to analyze the tradeoffs between location and routing factors at the same level of decision hierarchy. Among the exact solution methods developed for the LRP is Laporte et al.'s (1988) method which transforms the Multi-Depot Vehicle Routing Problem (MDVRP) and LRP into a constrained assignment problem solved by branch and bound. Ambrosino and Scutellà (2005) attribute the significance of strategic decisions like facility locations, transportation and inventory levels in the distribution network design problem (DNDP) to an article by Crainic and Laporte (1997), along with an extensive literature review of the LRP and DNDP. Ambrosino and Scutellà adopt Laporte's (1988) classification of LRPs. Their proposed integrated DNDP can be formulated as an LRP of category 4/R/T/T involving facility, warehousing and transportation as well as inventory decisions. This category label means distribution networks are made up of four layers, with routes of type replenishment (direct shipments) and type tour (vehicle routes). Jacobsen and Madsen (1980) model a newspaper delivery system as a three layer LRP and suggest three heuristic methods for the problem. Another study which introduces three heuristics for the LRP is due to Srivastava (1993). Tüzün and Burke (1999) propose a two-phase tabu search architecture for the solution of the standard 2-layer multi-depot LRP where depots have unlimited throughput capacity. Wu et al. (2002) decompose the standard LRP with capacitated depots into a facility location-allocation problem and a vehicle routing problem, and try then to solve both subproblems using simulated annealing. For the same class of the LRP, Albareda-Sambola et al. (2005) apply a method that generates first a lower bound either from the linear relaxation of the given problem or from the solutions of a pair of ad hoc knapsack and asymmetric traveling salesman problems. This lower bound is then used as a starting point of a tabu search heuristic. Lastly, Melechovský et al. (2005) address an LRP with nonlinear depot costs that grow with the total demand satisfied by the depots. They present a hybrid metaheuristic method consisting of tabu search and variable neighborhood search heuristics. We are aware of one study by Aksen and Altinkemer (2005) on Lagrangian relaxation for the LRP. They propose a 3-layer distribution logistics model for the conversion from brick-and-mortar to click-and-mortar retailing. A static one-period optimization model is built and solved using Lagrangian relaxation.

In this paper, we solve a 2-layer multi-depot location-routing problem (MDLRP) where transportation follows directly from depots to customers. There exist two kinds of depots: present depots and candidate depots. Present depots are already operating facilities that can be preserved or closed. If a present depot is closed, a fixed closing cost is incurred. This cost may turn out to be a gain since the closure of a depot usually brings about savings in overhead costs. Candidate depot locations are potential sites in which new depots can be opened. For each new depot to be opened, a fixed opening cost is incurred. In addition, there exist fixed operating costs which are charged for each preserved or newly opened depot. Customers are visited by a homogenous fleet of capacitated vehicles. For each of them, a vehicle acquisition cost is charged. Each customer has a deterministic demand which should be satisfied by the single visit of a vehicle. There is no capacity constraint on depots. The sum of depot opening-closing and operating, vehicle acquisition and traveling costs is minimized subject to the vehicle capacity in the problem.

The remainder of the paper is organized as follows. The problem description and its mathematical model are given in Section 2. In Section 3, detailed explanations of the Lagrangian relaxation scheme and the solution methods for the subproblems are provided. The heuristic method that is used to obtain upper bounds on the true optimal solution is detailed in Section 4. Section 5 presents the computational experiments, results and comparisons. Finally, Section 6 comprises a summary with concluding remarks.

2. MATHEMATICAL MODEL FOR THE MDLRP

In the MDLRP, the sum of depot opening-closing, vehicle acquisition and traveling costs is minimized subject to vehicle capacity constraints. According to Laporte's (1988) classification of LRPs, our problem is 2/T. It means there exist 2 layers; namely, depots and customers where the transportation between these layers is realized via tours (vehicle routes). The objective function and constraints of the model can be stated as follows:

$$\mathbf{P}: \text{ Min} \sum_{k\in ID} OC_k y_k + \sum_{k\in ID_{cand}} FC_k y_k + \sum_{k\in ID_{pres}} FC_k(1-y_k) + \sum_{i\in ID}\sum_{j\in IC} VC_i x_{iji}$$

$$+ \sum_{k\in ID}\sum_{i\in I}\sum_{\substack{j\in I\\ i\neq j}} c_{ij} x_{ijk} \tag{1}$$

$$\text{s.t.: } \sum_{k\in ID} \delta_{ik} = 1 \qquad \forall i\in IC \tag{2}$$

$$\sum_{\substack{j\in IC\cup\{k\}\\ j\neq i}} x_{ijk} = \delta_{ik} \qquad \forall i\in IC,\ \forall k\in ID \tag{3}$$

$$\sum_{\substack{j\in IC\cup\{k\}\\ j\neq i}} x_{jik} = \delta_{ik} \qquad \forall i\in IC,\ \forall k\in ID \tag{4}$$

$$\sum_{i\in IC} x_{ikk} = \sum_{i\in IC} x_{kik} \qquad \forall i\in ID, \forall k\in ID \tag{5}$$

$$\sum_{k\in ID}\sum_{i\in IC} x_{kik} + \sum_{k\in ID}\sum_{i\in IC}\sum_{\substack{j\in IC\\ i\neq j}} x_{ijk} = |IC| \tag{6}$$

$$\sum_{k\in ID}\sum_{i\in S}\sum_{\substack{j\in S\\ j\neq i}} x_{ijk} \leq |S| - L(S) \qquad \forall S\subseteq IC,\ |S|\geq 2 \tag{7}$$

$$\sum_{i\in IC} \delta_{ik} \leq |IC|\, y_k \qquad \forall k\in ID \tag{8}$$

$$x_{ijk} \in \{0,1\} \qquad \forall i\in I, \forall j\in I, \forall k\in ID \tag{9}$$

$$\delta_{ik}, y_k \in \{0,1\} \qquad \forall i\in I, \forall k\in ID \tag{10}$$

IC, *ID* and *I* in the model denote the set of customer nodes, the set of depot nodes and the union of these two sets, respectively. *ID* consists of ID_{pres} and ID_{cand}. The former is the set of already existing depots, and the latter is the set of candidate depots that can be opened. There are three sets of binary decision variables: x_{ijk} is equal to 1 if node j is visited after node i on a route originating from depot k. The variable y_k is equal to 1 if depot k is opened for $k \in ID_{cand}$, or if it is preserved for $k \in ID_{pres}$. The binary variable δ_{ik} is equal to 1 if customer i is assigned to depot k. FC_k is the fixed cost of having depot k in the solution. If depot k is not already present, then FC_k will have a relatively large positive value. Otherwise, it will denote the cost (gain) of closing depot k and will possibly have a negative value. OC_k is the depot operating cost. VC_k denotes the unit vehicle acquisition cost at depot k. Parameter c_{ij} denotes the traveling cost of one vehicle from node i to node j. M denotes a big number and Q is the uniform vehicle capacity. $L(S)$ in Eq.(7) is the optimal solution to the one-dimensional bin packing problem where the bin length is equal to the vehicle capacity Q, and demand values d_i ($i \in IC$) are the sizes of items to be packed into the bins.

The objective of **P**, shown in Eq. (1), is a combination of objectives of a facility location-allocation problem (FLAP) and a multi-depot vehicle routing problem (MDVRP). The constraints are comprised of pure FLAP constraints, pure MDVRP constraints and coupling constraints linking routing decisions with location decisions. Equation (2) assigns each customer to a depot. Equations (3)–(4) are flow conservation constraints which ensure all customers be visited exactly once on a route originating from the assigned depot. Equation (5) ensures that the numbers of incoming and outgoing arcs at each depot are equal. Equation (6) is identical to the sum of the constraints in Eq. (4). In order to obtain the second subproblem as a minimum spanning forest like problem after the Lagrangian relaxation, we add this redundant constraint to the model. Equation (7) is the well-known subtour elimination constraints which ensure all routes will start and end at a depot. The assignment of a customer to a closed or unopened depot, and routes originating from such a depot are avoided by Eq. (8). Finally, Eqs. (9)–(10) are integrality constraints.

3. LAGRANGIAN RELAXATION FOR THE MDLRP

Lagrangian relaxation is a decomposition method used for a variety of NP-hard optimization problems (see Geoffrion 1974). In this method, the true optimal objective value of the problem (Z^*_P) is bracketed between a lower and an upper bound [Z_{lb}, Z_{ub}]. In case of minimization, a good feasible solution constitutes an upper bound for the optimal solution of the minimization problem while its lower bound is obtained by solving the Lagrangian relaxed problem. The quality of the solution is assessed based on the gap between these two bounds. When Lagrangian relaxation is applied to the MDLRP in Eqs. (1)–(10), the coupling constraints in Eqs. (3)–(4) are relaxed. The left-hand sides of the constraints are subtracted from their right-hand sides. The differences are multiplied by the Lagrange multipliers λ and μ, respectively, which are unrestricted in sign. The terms are then augmented into the original objective function and the new objective function $ZLR(\lambda, \mu)$ in Eq. (11) is obtained.

$$ZLR(\lambda,\mu) = \sum_{k\in ID} OC_k y_k + \sum_{k\in ID_{cand}} FC_k y_k + \sum_{k\in ID_{pres}} FC_k (1-y_k) + \sum_{i\in ID} \sum_{j\in IC} VC_i x_{iji}$$
$$+ \sum_{k\in ID} \sum_{i\in I} \sum_{\substack{j\in I \\ j\neq i}} c_{ij} x_{ijk} + \sum_{k\in ID} \sum_{i\in IC} \sum_{\substack{j\in IC\cup\{k\} \\ j\neq i}} (\delta_{ik} - x_{ijk})\lambda_{ik}$$
$$+ \sum_{k\in ID} \sum_{j\in IC} \sum_{\substack{i\in IC\cup\{k\} \\ i\neq j}} (\delta_{jk} - x_{ijk})\mu_{jk} \qquad (11)$$

The resulting Lagrangian relaxed problem **LR**(λ, μ) can be partitioned into two independent subproblems. The first subproblem resembles an uncapacitated FLAP (**SubP1**) and can be solved with Cplex 9.0 in reasonable time. The second one is similar to a degree constrained minimum spanning forest problem (DCMSF) (**SubP2**). However, **SubP2** is still an NP-hard problem, which is tackled with an augmented Lagrangian relaxation by relaxing the degree constraints. The relaxed **SubP2** becomes a minimum spanning forest problem with a minimum number of outgoing arcs at root nodes (depots). It is solved with a modified version of Prim's minimum spanning tree algorithm. Figure 1 displays the flow chart of the iterative subgradient optimization procedure with the Lagrangian relaxation scheme applied to the parent problem **P**. The flow chart's segment in the box shows the inner augmented Lagrangian relaxation which is applied to the second subproblem **SubP2**. The structure of the Lagrangian relaxed problem LR(λ, μ) is presented below in plain English.

LR(λ, μ): Minimize *ZLR*(λ, μ) = Σ*Augmented* FLAP *objectives*
+ Σ*Augmented* MDVRP *objectives*

subject to: i. Pure FLAP constraints (2)–(8)
ii. Pure MDVRP constraints (5)–(7)
iii. Nonnegativity and integrality constraints (9)–(10)

3.1 The Lagrangian relaxed problem LR

ZLR(λ, μ) which is the objective function of the Lagrangian relaxed problem turns out to be separable into two as FLAP and MDVRP objectives. In order to obtain two independent components, *ZLR*(λ, μ) needs to be rearranged. One part of the relaxed constraints that are augmented into the objective function can be separated as shown in Eq. (12). The Lagrange multipliers in this FLAP objective component represent pseudo costs of allocating customers to depots.

$$\sum_{i\in IC} \sum_{k\in ID} \delta_{ik} (\mu_{ik} + \lambda_{ik}) \qquad (12)$$

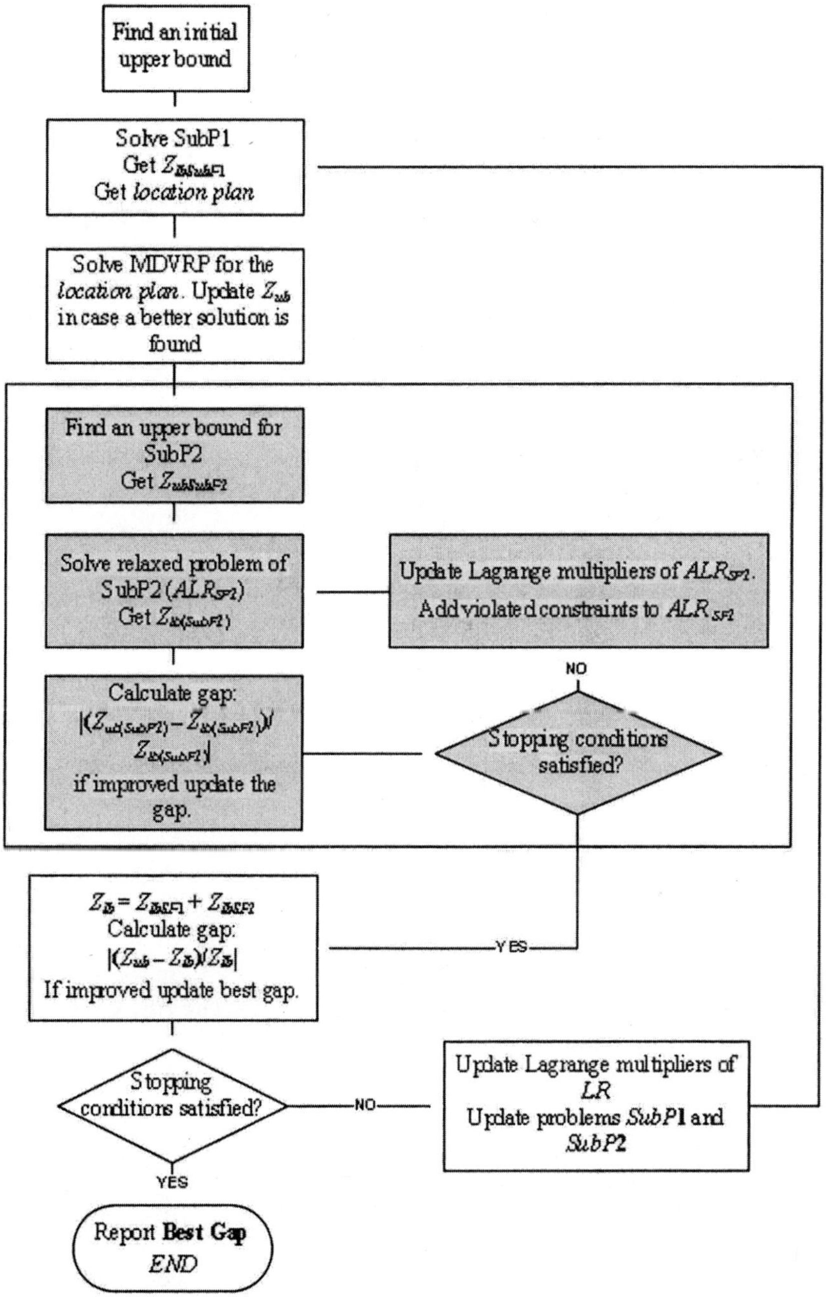
Find an initial upper bound
Solve SubP1
Get $Z_{lb(SubP1)}$
Get *location plan*
Solve MDVRP for the *location plan*. Update Z_{ub} in case a better solution is found
Find an upper bound for SubP2
Get $Z_{ub(SubP2)}$
Solve relaxed problem of SubP2 (ALR_{SP2})
Get $Z_{lb(SubP2)}$
Update Lagrange multipliers of ALR_{SP2}.
Add violated constraints to ALR_{SP2}
NO
Calculate gap:
$|(Z_{ub(SubP2)} - Z_{lb(SubP2)})/Z_{lb(SubP2)}|$
if improved update the gap.
Stopping conditions satisfied?
$Z_{lb} = Z_{lb(SP1)} + Z_{lb(SP2)}$
Calculate gap:
$|(Z_{ub} - Z_{lb})/Z_{lb}|$
If improved update best gap.
YES
Stopping conditions satisfied?
NO
Update Lagrange multipliers of *LR*
Update problems *SubP1* and *SubP2*
YES
Report **Best Gap**
END

By reordering the remaining terms in Eq. (11), we derive the 3-dimensional asymmetric and depot dependent traveling cost matrix $C_{new}=[(c_{ijk})^{new}]$. In this cost structure, the cost of traveling from node i to node j by a vehicle not only depends on the distance between i and j, but also on the depot k which sends off that vehicle. Let $G(I,A)$ denote the complete weighted and directed graph of customers and depots, i.e. $A = \{(i,j) \in (I \times I), i ? j\}$. Let $(c_{ijk})^{new}$ denote the cost of arc (i, j), if it is traversed with a vehicle dispatched from depot $k \in ID$. Arc costs in G are then defined as follows:

1. $(i,j) \in IC \times IC, i ? j, k \in ID : (c_{ijk})^{new} = c_{ij} - \lambda_{ik} - \mu_{jk}$
2. $(i,j) \in IC \times ID \qquad : (c_{ijj})^{new} = c_{ij} - \lambda_{ij}$
3. $(i,j) \in ID \times IC \qquad : (c_{iji})^{new} = c_{ij} - \mu_{ji} + VC_i$
4. $(i,j) \in ID \times IC, k \in ID, i ? k : (c_{ijk})^{new} = +8$
5. $(i,j) \in IC \times ID, k \in ID, j ? k : (c_{ijk})^{new} = +8$

The last two cost assignments avoid illegal arc definitions. An arc that is emanating from a depot or entering a depot cannot be defined on a route which originates from a different depot. That means, if i or j is a depot node and k is another depot node, than x_{ijk} cannot be 1. For this, c_{ijk} is assigned to infinity in these cases. After the relevant rearrangements, the Lagrangian relaxed problem **LR**(λ, μ) can be stated as follows.

$$\text{Min} \quad ZLR(\lambda,\mu) = \sum_{k\in ID} OC_k y_k + \sum_{k\in ID_{cand}} FC_k y_k + \sum_{k\in ID_{pres}} FC_k (1-y_k) + \sum_{k\in ID}\sum_{i\in IC} \delta_{ik}(\lambda_{ik} + \mu_{ik}) + \sum_{k\in ID}\sum_{i\in I}\sum_{\substack{j\in I\\ j\neq i}} (c_{ijk})^{new} x_{ijk} \tag{13}$$

subject to : (2), (5)–(10)

We use a tabu search heuristic to find a good feasible solution whose objective value will be an upper bound on Z^*_P, the true optimal objective value of the problem **P**. This upper bound is updated throughout the subgradient iterations of the Lagrangian relaxation. The upper bound generation and updating method are explained in Section 4.

3.2 Subgradient optimization

Let SG^q denote the subgradient vector of the problem LR(λ, μ) at iteration q of the subgradient optimization procedure. Step size s^q is then derived from the norm square of SG^q and the gap between the current best objective Z_{ub} (upper bound on Z^*_P) and current Lagrangian objective Z^q_{LR}. It is multiplied also by a scalar $?_q$ whose first value $?_1$ is 2.0 by convention (see Fisher, 1981). This scalar is halved whenever the objective Z_{LR} does not improve for a specified number of consecutive iterations. At the beginning, we set all Lagrange multipliers to the initial value zero.

Formulae of the subgradient optimization routine for the Lagrangian relaxation of problem **P** are given below.

$$(SG_{ik}^{\lambda})^q = (\delta_{ik})^q - \sum_{\substack{j\in IC\cup\{k\} \\ j\neq i}} (x_{ijk})^q \qquad \forall i\in IC, \forall k\in ID \tag{14}$$

$$(SG_{ik}^{\mu})^q = (\delta_{ik})^q - \sum_{\substack{j\in IC\cup\{k\} \\ j\neq i}} (x_{jik})^q \qquad \forall i\in IC, \forall k\in ID \tag{15}$$

$$\|SG\|^2 = \|SG^{\lambda}\|^2 + \|SG^{\mu}\|^2 \qquad s^q = \Lambda^q \frac{Z_{ub}^q - Z_{LR}^q(\lambda,\mu)}{\|SG\|^2}$$

$$(\lambda_{ik})^{q+1} = (\lambda_{ik})^q + s^q (SG_{ik})^q \qquad \forall i\in IC, \forall k\in ID \tag{16}$$

$$(\mu_{ik})^{q+1} = (\mu_{ik})^q + s^q (SG_{ik})^q \qquad \forall i\in IC, \forall k\in ID \tag{17}$$

3.3 FLAP-like problem SubP1

The first of the two subproblems comprising **LR**(λ, μ) is the FLAP like problem **SubP1**. The formulation of **SubP1** can be written as follows.

$$\text{Min } Z_{SubP1} = \sum_{k\in ID} OC_k y_k + \sum_{k\in ID_{cand}} FC_k y_k + \sum_{k\in ID_{pres}} FC_k (1-y_k) + \sum_{k\in ID}\sum_{i\in IC} \delta_{ik}(\lambda_{ik}+\mu_{ik}) \tag{18}$$

subject to : (2), (8), (10)

Since the technological coefficients matrix of **SubP1** is unimodular, we can define δ_{ik}'s as positive continuous variables between 0 and 1 instead of binary variables. Furthermore, the constraints in (8) should be disaggregated as $\delta_{ik} \le y_k \quad \forall i\in IC,\ \forall k\in ID$. As a result, we get a formulation with less integer variables, which in turn yields significantly better solution times for **SubP1**. The depots to be preserved or opened are determined, and customers are allocated to these. Each depot has its operating (OC_k) and opening cost (FC_k). FC_k is incurred when a candidate depot is opened or when a present depot is closed. FC_k of the present depots is generally negative corresponding to the savings in overhead costs and to the salvage value obtained by closing the depot. The Lagrange multipliers λ and μ represent allocation costs of customers to depots. These costs change as the Lagrange multipliers get updated during the subgradient optimization iterations on the

Lagrangian relaxed problem. At the beginning of each subgradient iteration, allocation costs are plugged in and **SubP1** is solved with Cplex to optimality. The solution times are generally reasonable. A problem instance with 20 depots and 1000 customers takes 2.84 seconds on a present-day desktop PC.

3.4 Minimum spanning forest-like subproblem SubP2

The second subproblem of **LR**(λ, μ) is **SubP2** which resembles a degree and capacity constrained minimum spanning forest problem (CMSF). The cost matrix C_{new} comprises the coefficients in the objective function of the subproblem. Since the Lagrange multipliers λ and μ are embedded in this matrix, C_{new} changes as the multipliers change at each subgradient iteration. The mathematical formulation of **SubP2** can be stated as follows:

$$\text{Min } Z_{SubP2} = \sum_{k \in ID} \sum_{i \in I} \sum_{\substack{j \in I \\ j \neq i}} (c_{ijk})^{new} x_{ijk} \tag{19}$$

subject to : (5)–(7), (9)

In **SubP2** depot locations $k \in ID$ represent center nodes, while customers $i \in IC$ are terminal nodes. The terminals should be accessible from one of the center nodes via a subtree rooted at that center. Eq. (5) enforces that the numbers of outgoing and incoming arcs at each center node be equal. This balance-of-in-and-outdegree condition differentiates **SubP2** from the classical MSF. Capacity and subtour elimination constraints are given in Eq. (7). The capacity constraint requires that the total demand on a subtree rooted at a center node do not exceed Q. Equation (6) provides connectivity of the tree, while Eq. (7) avoids the formation of subtrees which are not linked to any of the center nodes. Since the constraints in Eqs. (3)–(4) are relaxed, any node can have more than one offspring nodes.

SubP2 is actually still hard to solve. If balance of degree constraints are discarded, and if the number of depots in ID is dropped to one, **SubP2** would reduce to the capacitated minimum spanning tree (CMST) problem. Papadimitriou (1978) showed that CMST is an NP-hard problem. Hence, **SubP2** also belongs to the NP-hard class. In order to solve **SubP2** we use the method proposed in Aksen and Altinkemer (2005) where the augmented Lagrangian relaxation method of Gavish (1985) is adopted and modified to tackle balance of degree constraints. We relax the subtour elimination constraints in **SubP2**, since this relaxation scheme achieves empirically better lower bounds on Z^*_{SubP2}. First, the constraint set in Eq. (7) is divided into two parts as (7.a) and (7.b), the second of which is relaxed. Secondly, a trivial constraint which sets the minimum number of vehicles required is added to the original formulation as (7.c). This minimum number is calculated by solving the associated bin-packing problem that embraces all demand values d_i, $i \in IC$.

$$\sum_{k \in ID} \sum_{i \in S} \sum_{\substack{j \in S \\ j \neq i}} x_{ijk} \leq |S| - 1 \qquad \forall S \subseteq IC \ \exists |S| \geq 2 \qquad (7.a)$$

$$\sum_{k \in ID} \sum_{i \in S} \sum_{j \in S} x_{ijk} \leq |S| - \mathrm{L}_S \qquad \forall S \subseteq IC, |S| \geq 2 \qquad (7.b)$$

$$\sum_{k \in ID} \sum_{i \in IC} x_{kik} \geq \mathrm{L}_{IC} \qquad (7.c)$$

The relaxed constraint set (7.b) is multiplied with Lagrange multipliers α, where $\alpha = 0$. Left hand side values are subtracted from their right hand sides and the resulting terms are augmented into the objective function of **SubP2** in Eq. (19). In order to combine the embedded terms with Z_{SubP2} and to get a compact formulation for the objective function of the problem after the Lagrangian relaxation we separate Eq. (19) into three parts as follows:

$$\sum_{k \in ID} \sum_{i \in I} \sum_{\substack{j \in I \\ j \neq i}} (c_{ijk})_{new} x_{ijk} = \sum_{k \in ID} \sum_{i \in IC} (c_{kik})^{new} x_{kik} + \sum_{k \in ID} \sum_{i \in IC} (c_{ikk})^{new} x_{ikk} + \sum_{k \in ID} \sum_{i \in IC} \sum_{\substack{j \in IC \\ j \neq i}} (c_{ijk})^{new} x_{ijk} \qquad (20)$$

After necessary rearrangements, the objective function and constraints of **ALR**^{SubP2} (the Lagrangian relaxed **SubP2**) can be stated as follows:

$$\text{Min } Z_{LR}^{SubP2}(\alpha) = \sum_{k \in ID} \sum_{i \in IC} \left[(c_{kik})^{new} \right] x_{kik} + \sum_{k \in ID} \sum_{i \in IC} \left[(c_{ikk})^{new} \right] x_{ikk} + \sum_{k \in ID} \sum_{i \in IC} \sum_{\substack{j \in IC \\ j \neq i}} \left[(c_{ijk})^{new} - \sum_{s \in G_i} \alpha_s \right] x_{ijk} + \sum_{s \in ?} (|S| - \mathrm{L}_S) \alpha_s \qquad (21)$$

subject to : (5), (6), (7.a), (7.c), (9)

The last term in Eq. (21) is constant for a given set of Lagrange multipliers α. Since the solution to **ALR**^{SubP2} will constitute a lower bound for the optimal solution of **LR**, omitting of the constant term would overestimate or underestimate the lower bound depending on the negativity of the terms. Observe that S in the relaxed constraints represents any unordered subset of IC with a cardinality greater than one, which requires two or more vehicles to deliver orders. The set of such subsets is denoted by $\mathbf{\Psi}$. For each $S \in \mathbf{\Psi}$, there is an associated Lagrange multiplier $\alpha_s = 0$. Let G_{ij} denote the index set of subsets S in $\mathbf{\Psi}$ that contain customer nodes i and j. The augmented Lagrangian relaxation feature is used here, because we do not explicitly generate all constraints in Eq. (7.b). Therefore, we do not compute the entire multiplier vector α, either. The augmented Lagrangian relaxed problem $\mathbf{ALR}^{SubP2}$ is equivalent to an MSF problem without capacity constraints where the cost matrix C_{new} is dependent on the center node of departure. However, there are two distinct restrictions in this MSF problem:

- The sum of outgoing degrees of all center nodes has to be equal to or greater than L_{IC} as required by the constraints in Eq. (7.c).
- At each center node, incoming and outgoing degrees should be equal as required by constraints in Eq. (5).

The solution of the problem $\mathbf{ALR}^{SubP2}$ is checked against the violation of constraints in Eq. (7.b) in **SubP2**. If any violated constraint is detected, it is added together with its associated Lagrange multiplier to the set of active constraints and multipliers. The objective function is augmented with the product of the difference between the violated constraint's right- and left-hand side values and the associated Lagrange multiplier's initial value. We do not remove previously augmented constraints from the set of active constraints in the Lagrangian problem; neither do we generate any such constraint for a second time. Gavish explains a further technique to generate a tight Lagrangian objective function by finding an initial multiplier value for every augmented constraint while maintaining the optimality property of the Lagrangian solution before that constraint. We adopted this technique into our augmented Lagrangian relaxation of **SubP2**. Finally, the degree balance constraints in Eq. (5) and the minimum sum constraint in Eq. (7.c) on the center nodes' outgoing degrees should be reckoned with.

The closest version of $\mathbf{ALR}^{SubP2}$ is the degree-constrained minimum spanning tree problem (DCMST). Garey and Johnson (1979) prove that the DCMST with arbitrary degree constraints on nodes other than the center is NP-hard. In spite of copious methods and algorithms developed for the DCMST in the literature, we cannot use any of them as is. First of all, $\mathbf{ALR}^{SubP2}$ displays a forest structure with asymmetrical and center-node dependent costs. Secondly, the degree constraints that appear in $\mathbf{ALR}^{SubP2}$ relate to the balance of incoming and outgoing degrees at the center nodes only. There exists also a lower bound on the sum of outgoing degrees at those centers. From this perspective, $\mathbf{ALR}^{SubP2}$ is conceivably easier to solve than a general DCMST problem. Aksen and Altinkemer (2005) develop a polynomial-time procedure called [MSF-ALR] which is largely an adaptation of Prim's MST algorithm. We take on their solution method for solving the problem $\mathbf{ALR}^{SubP2}$.

3.5 Subgradient optimization in the augmented Lagrangian relaxation

The subgradient vector $\mathbf{Y}$ is calculated according to the formulae given below. The cardinality of the subgradient vector increases as the number of violated constraints goes up. In the formulae, G^q denotes the index set of those subtour elimination and capacity constraints in Eq. (7.b) which have been violated and thus generated either in the current iteration q or in a previous iteration. Each index r in G^q corresponds to some subtree of customer nodes whose indices comprise a particular subset S in $\mathbf{\Psi}$ as explained in Section 3.4. There are as many as $|G^q|$ constraints from Eq. (7.b) relaxed and augmented into $\mathbf{ALR}^{SubP2}$. In Eq. (23), s^q_{ALR} denotes the step size of the subgradient optimization, ? ${}^{q}_{ALR}$ is a scalar with the initial value 2.0, $Z^q_{ub(SubP2)}$ is an upper bound on the true optimal objective value of **SubP2**, and finally $Z^q_{ALR(SubP2)}$ is the current augmented Lagrangian objective value. The scalar ? $_{ALR}$ is halved whenever $Z^q_{ALR(SubP2)}$ does not increase for a specified number of consecutive iterations. S_r in Eq. (22) indicates the rth subset of customers in $\mathbf{\Psi}$ which are spanned by the same subtree.

$$(\Upsilon_r)^q = \left(|S_r| - \mathrm{L}_{S_r}\right) - \sum_{k \in ID} \sum_{i \in S_r} \sum_{\substack{j \in S_r \\ j \neq i}} \left(x_{ijk}\right)^q \qquad \forall r \in G^q \tag{22}$$

$${}^{q}_{ALR} = \Lambda^q_{ALR} \frac{Z^q_{ub(SubP2)} \quad Z^q_{ALR(SubP2)}(\alpha)}{\|\Upsilon\|^2} \tag{23}$$

$$(\alpha_r)^{q+1} = \min\left\{0, (\alpha_r)^q + s^q_{ALR}\left(\Upsilon^{\alpha}_r\right)^q\right\} \qquad \forall r \in G^q \tag{24}$$

4. GENERATING UPPER BOUNDS FOR P

At each subgradient iteration of the outer Lagrangian relaxation of **P**, the solution obtained for **SubP1** reveals which depots are preserved and which ones are opened. Once this information is provided, the remainder of the problem becomes a MDVRP any feasible solution of which constitutes an upper bound to **P**. Each time a new depot location plan is obtained by solving **SubP1**, a tabu search (TS) heuristic is triggered in the hope of achieving a better upper bound for **P**. When the Lagrangian iterations terminate, a greedy method called Add-Drop heuristic starts in case the final gap is greater than 2%. First, closed or unopened depots are added to the solution one by one; then, currently open depots are dropped from the solution in a similar decremental fashion. An MDVRP is solved with respect to each of these scenarios. If a better feasible solution is realized, the new depot location plan is adopted, and Z_{ub} is updated.

TS is an meta-heuristic algorithm that guides the local search to prevent it from being trapped in premature local optima or in cycling. It starts with an initial solution. At each iteration of the TS, a neighborhood of solutions is generated for the current solution. The best one from this neighborhood is picked as the current solution depending on a number of criteria. Certain attributes of previous solutions are kept in a tabu list which is updated at the end of each iteration. The selection of the best solution in the neighborhood is done such that it does not attain any of the tabu attributes. The best feasible solution so far (incumbent) is updated if the current solution is both feasible and better than the incumbent. The procedure continues until one or more stopping criteria are fulfilled. In our study, we adopted the same tabu search procedure as proposed by Aksen et al. (2006) for the open vehicle routing problem with fixed driver nodes. We tailored the procedure for the MDVRP, and also enriched it with additional neighborhood generation moves.

4.1 An initial solution for P

In order to generate an initial solution for our TS, we make use of the constructive heuristic [PFIH-NN] proposed by Aksen and Altinkemer (2003). It is a hybrid of Push Forward Insertion and Nearest Neighborhood methods where customers are first assigned to the nearest depot. They are placed in an array sorted in the non-decreasing order of a special cost coefficient. This coefficient is calculated for each customer based on his distance to the assigned depot. The customer with the lowest cost coefficient is appended to a route. The remaining customers in the array are then chosen one at a time, and inserted into this first route according to the cheapest insertion principle. When the next to-be-inserted customer's demand exceeds the spare capacity on the current route, a new route is initiated.

4.2 Evaluation of solutions

For a given location plan the objective of the problem is to minimize the vehicle acquisition and total traveling cost. In our tabu search method, we apply strategic oscillation by admitting infeasible solutions where infeasible solutions are penalized in proportion to the violation of capacity constraints. The penalty terms are added to the objective value of an infeasible solution. Penalty coefficients are updated every 10 iterations based on the number of feasible and infeasible solutions visited. The objective value for a solution is given by $\sum_{k\in I}\sum_{D\notin I}\sum_{\substack{\in I\\ j\neq i}} c_{ijk} + \sum_{r\in R} p_c V_c(r)$ where the first term is the total traveling cost, R is the set of all routes, $V_c(r)$ denotes the overload (total demand of customers in route r minus vehicle capacity Q), and p_c denotes the penalty coefficient for overload on a route.

4.3 Neighborhood structure and tabu attributes

We use four move operators to create a neighborhood for the current solution. A pictorial description of the first three can be found in the paper by Tarantilis and Kiranoudis (2002). Each move involves two pilot nodes:

1-0 *move* : One of the pilot nodes is taken from its current position and inserted after the other.

1-1 *exchange* : Two pilot nodes are swapped.

2-*Opt move* : For two pilot nodes in the same route, the arcs emanating from these are removed. Two arcs are added one of which connects the pilot nodes, and the other connects their successor nodes. If the pilot nodes are in different routes, then the route segments following them are swapped preserving the order of nodes succeeding the pilots on each segment.

2-2 *exchange* : One of the pilot nodes and its successor are swapped with the other pilot node and its successor.

The size of the neighborhood generated in each iteration depends on the number of operating depots and the number of customer nodes in the problem. Besides neighborhood generation, we incorporate also a local search with these moves into the tabu search as a tool of local post optimization (LPO). A series of LPO operations are to be applied to the initial solution, to the current solution at the end of every 100 iterations if it is feasible, and also to the incumbent (current best solution) whenever it is updated. This helps the intensification of tabu search on the given MDVRP instance. We determine the sequence of LPO operations empirically, according to the results of extensive experimentation. In the application of LPO, all customers are set one by one as the first pilot node. For a given pilot node, the second one is chosen such that the related move yields the highest improvement in total distance without causing any infeasibility.

The tabu list is updated at the end of each iteration. Tabu attributes of a solution generated by a move can be stated as follows.

1-0 *move* : If node i is inserted after node j, the position of i cannot be changed by the same move while it is tabu-active.

1-1 *exchange* : If nodes i and j are swapped, they cannot be swapped again while they are tabu-active.

2-*Opt move* : If 2-Opt move is applied to nodes i and j, the move cannot be applied again to the same nodes while they are tabu-active.

2-2 Exchange: If nodes i and $(i+1)$ are swapped with nodes j and $(j+1)$, these cannot be swapped again while they are tabu active.

At each iteration, the tabu tenure is selected randomly between 5 and 15 iterations. In some cases, namely if the so-called aspiration criterion is satisfied, a move can be executed although its attributes are tabu-active. Aspiration criterion is considered to be satisfied if the total distance resulting from the move is better than the incumbent's objective value.

4.4 Stopping criteria

Tabu search terminates when any one of two stopping criteria is satisfied. The first criterion is the total number of iterations performed. The second criterion is the maximum permissible number of iterations during which the best feasible or best infeasible solution does not improve. Both values are determined based on the number of customers and on the number of operating depots found in the solution of **SubP1**.

5. COMPUTATIONAL RESULTS

The code of the proposed method is written in ANSI C language, compiled in Visual C++ .NET and executed on a 3.20 GHz Intel Xeon processor with 2 GB RAM. The algorithm is tested with 44 problems which consist of two parts. The first part includes 20 randomly generated small size test problems with 15 up to 35 customers and 2 up to 6 depots. The second part comprises 24 problems solved in Tüzün and Burke (1999) which have 100 up to 150 customers and 10 up to 20 depots. The problems in the first part are also solved by Cplex 9.0 with a time limit of five hours. These small size problems constitute benchmarks for the upper bounds obtained by our method. Upper bounds to the problems in the second part are compared with the solutions found in Tüzün and Burke (1999).

The stopping conditions of the Lagrangian relaxation have been fine-tuned by extensive experimentation on 16 test problems. Since the solution times of the larger problems are not practical for such experimentation, 10 of these problems have been selected from the ones in the first part. The mutually exclusive stopping conditions of the subgradient optimization for the outer Lagrangian relaxation are fixed as follows. If the number of subgradient iterations performed exceeds 300, or if the number of consecutive subgradient iterations during which the Lagrangian gap does not improve reaches 100, or finally if the amount of absolute increment in the Lagrange multipliers is not greater than 1.0e–7, the subgradient optimization procedure for the problem **P** stops. The stopping conditions in case of the augmented Lagrangian relaxation applied to **SubP2** are satisfied if the predefined limit on one of following parameters is reached: 150 subgradient iterations performed in the augmented Lagrangian relaxation, the step size or the gap between $Z_{lb(SubP2)}$ and $Z_{ub(SubP2)}$ dropping below 1.0e–5, and finally 75 consecutive iterations during which the gap does not improve.

Table 1. Results for 20 randomly generated test problems with *nc* between 15 and 35

Nc	*npd*	*ncd*	Z_{Cplex}	%GAP2	Z_{lb}	Z_{ub}	%GAP1	CPU(s)
15	1	2	1127.84	0.00%	1075.58	1127.84	4.86%	54.69
15	1	2	994.92	0.00%	994.92	994.92	0.00%	27.56
15	-	2	1024.19	0.29%	975.28	1027.14	5.32%	53.49
15	-	2	1032.08	4.13%	1031.51	1074.68	4.19%	48.46
20	1	3	1136.52	1.07%	1128.51	1148.72	1.79%	38.15
20	1	3	1285.05	2.56%	1262.14	1317.96	4.42%	96.66
20	-	3	1442.47	0.00%	1435.11	1442.48	0.51%	155.10
20	-	3	1022.49	0.00%	953.04	1022.49	7.29%	74.44
25	1	2	1407.29	−0.34%	1321.13	1402.44	6.15%	261.56
25	1	2	1271.85	1.18%	1244.53	1286.85	3.40%	161.76
25	1	4	1424.57	−0.45%	1370.51	1418.18	3.48%	241.80
25	1	4	1368.62	0.14%	1367.03	1370.47	0.25%	241.31
30	1	4	1629.90	−6.43%	1471.51	1525.03	3.64%	356.78
30	1	4	1432.56	0.65%	1348.73	1441.94	6.91%	640.78
30	-	4	1599.46	−3.48%	1511.43	1543.86	2.15%	232.63
30	-	4	1619.42	−0.47%	1555.81	1611.87	3.60%	482.01
35	1	4	1909.93	−5.08%	1735.65	1812.81	4.45%	945.08
35	1	4	1408.74	−1.61%	1362.82	1386.02	1.70%	285.20
35	-	6	1844.70	−2.35%	1658.61	1801.37	8.61%	682.30
35	-	6	1730.64	−5.55%	1556.13	1634.60	5.04%	582.43
Averages			**1385.66**	**−0.79%**	**1318.00**	**1369.58**	**3.89%**	**283.11**

Table 2. Results for Tüzün and Burke's instances

prob id	*nc*	*ncd*	Z_{Cplex}	%GAP3	Z_{lb}	Z_{ub}	%GAP1	CPU(s)
P111112	100	10	1556.64	−8.95%	1283.09	1417.30	10.46%	19875.27
P111122	100	20	1531.88	−7.95%	1178.19	1410.04	19.68%	10554.93
P111212	100	10	1443.43	−2.57%	1140.54	1406.33	23.30%	9562.77
P111222	100	20	1511.39	−3.08%	1186.54	1464.84	23.45%	16420.19
P112112	100	10	1231.11	−1.72%	1079.16	1209.88	12.11%	14443.91
P112122	100	20	1132.02	−9.95%	925.16	1019.44	10.19%	18333.10
P112212	100	10	825.12	−11.95%	627.05	726.48	15.86%	7158.19
P112222	100	20	740.64	−0.31%	541.66	738.34	36.31%	15391.94
P113112	100	10	1316.98	−1.59%	1069.98	1296.04	21.13%	16432.57
P113122	100	20	1274.50	−8.98%	1055.33	1160.09	9.93%	12327.16
P113212	100	10	920.75	−1.30%	753.37	908.79	20.63%	6190.90
P113222	100	20	1042.21	−10.84%	780.93	929.22	18.99%	11696.95
P131112	150	10	2000.97	−6.57%	1561.25	1869.43	19.74%	52546.65
P131122	150	20	1892.84	0.35%	1465.80	1899.42	29.58%	54043.24
P131212	150	10	2022.11	3.83%	1589.11	2099.50	32.12%	43472.18
P131222	150	20	1854.97	−2.55%	1438.10	1807.63	25.70%	55900.30
P132112	150	10	1555.82	−4.34%	1151.67	1488.29	29.23%	42149.14
P132122	150	20	1478.80	1.58%	1144.07	1502.16	31.30%	59226.08
P132212	150	10	1231.34	0.26%	959.29	1234.50	28.69%	26122.60
P132222	150	20	948.28	−1.06%	742.16	938.22	26.42%	69757.69
P133112	150	10	1762.45	−5.38%	1232.78	1667.65	35.28%	10469.41
P133122	150	20	1488.34	−2.38%	1051.04	1452.97	38.24%	32540.27
P133212	150	10	1264.63	−7.22%	930.82	1173.29	26.05%	55394.52
P133222	150	20	1182.28	0.61%	973.35	1189.44	22.20%	26393.21
Averages			**1383.73**	**−3.84%**	**1077.52**	**1333.72**	**23.61%**	**28600.13**

For all of the small size problems, %GAP1 is under 10% and the average %GAP1 of these problems is 3.89%. For 10 out of the 20 problems Z_{ub} outperforms Z_{Cplex}, while for three of them the proposed method finds the same solution as Cplex. For seven of the problems Cplex does better than the proposed method; yet the maximum gap between Z_{ub} and Z_{Cplex} is 4.13%. The quality of %GAP1 diminishes in the problems of the second part. Although, there is no indication of a continuous increase in %GAP1 as the number of customers in the problem increases, we observe that the average %GAP1 of the problems with 150 customers is higher than that of the problems with 100 customers. The upper bounds found for Tüzün and Burke instances update 19 out of 24 solutions given in their study, while an average improvement of 3.84% is obtained. The solution times of the problems with more than 100 customers are significantly long which makes the revision of the implementation of the procedure imperative.

6. SUMMARY AND CONCLUSIONS

In this study, an uncapacitated multi-depot location routing problem (MDLRP) is solved using Lagrangian relaxation. Two subproblems emerge from the relaxation of the coupling constraints in the MDLRP model. The first of them has a structure similar to a facility location-allocation problem (FLAP), and is solved with Cplex 9.0 to optimality in negligible amount of time. The second one is a capacity and degree constrained minimum spanning forest-like problem which is still an NP-hard problem. To tackle it, an augmented Lagrangian relaxation is applied. The nested Lagrangian relaxation-based solution method is tested on 44 MDLRP instances which consist of 20 randomly generated problems and 24 problems solved in Tüzün and Burke (1999). For the problems in the first part, gaps are below 10%. In most of the small size problems, the final upper bounds are better than the corresponding Cplex solutions. For problems in the second part, gaps are higher with an average of 23.61%, while the upper bounds for these improve most of the solutions given in Tüzün and Burke (1999). The experimental results not only assess the performance of the proposed procedure, but also point to new research directions. The next step would be solving the MDLRP with time windows. This type of time restrictions is a crucial quality of service (QoS) guarantee promised more and more often to customers in distribution logistics. Finally, long solution times especially for problems with more than 100 customers are a critical disadvantage of the proposed method. This might be overcome by a new implementation of the modified Prim's algorithm which is used to solve the Lagrangian relaxed subproblem $\mathbf{ALR}^{SubP2}$.

ACKNOWLEDGMENTS

Deniz Aksen and Zeynep Özyurt have been supported by KÜMPEM (Koç University Migros Professional Training Center) for this research. The authors would like to thank the two anonymous referees for their insightful suggestions and comments which benefited the paper significantly.

REFERENCES

Ahipasaoglu, S.D., Erdogan, G. and Tansel, B., "Location-routing problems: a review and assessment of research directions", Working Paper IEOR 2003–07, Department of Industrial Engineering, Bilkent University, Ankara, Türkiye (2004).

Aksen, D. and Altinkemer, K., "Efficient frontier analysis and heuristics for etailing logistics", Working Paper, Purdue University, Krannert Graduate School of Management: West Lafayette, Indiana, USA (2003).

Aksen, D. and Altinkemer, K., "A location-routing problem for the conversion to the 'click-and-mortar' retailing: the static case", Working Paper, College of Administrative Science and Economics, Koç University, Istanbul, Türkiye (2005).

Aksen, D., Özyurt, Z. and Aras, N., "Open vehicle routing problem with driver nodes and time windows", *avaliable online in Journal of Operational Research Society,* August 2006, (doi:10.1057/palgrave.jors.2602249).

Albareda-Sambola, M., Díaz, J. A. and Fernández, E., "A compact model and tight bounds for a combined location-routing problem", *Computers & Operations Research* 32, 407–428 (2005).

Ambrosino, D. and Scutellà, M. G., "Distribution network design: new problems and related models", *European Journal of Operational Research* 165, 610–624 (2005).

Crainic, T. G. and Laporte, G., "Planning models for freight transportation", *European Journal of Operational Research* 97, 409–438 (1997).

Garey, G. and Johnson, D. S., "Computers and intractability: a guide to the theory of NP-completeness", W. H. Freeman and Company: New York (1979).

Gavish, B., "Augmented Lagrangian based algorithms for centralized network design", *IEEE Transactions on Communications* COM-33, 1247-1257 (1985).

Geoffrion, A. M., "Lagrangian relaxation and its uses in integer programming", *Mathematical Programming Study* 2, 82–114 (1974).

Jacobsen, S.K. and Madsen, O. B. G., "A comparative study of heuristics for a two-level routing-location problem", *European Journal of Operational Research* 5, 378–387 (1980).

Laporte, G., Nobert, Y. and Taillefer, S., "Solving a family of multi-depot vehicle routing and location-routing problems", *Transportation Science* 22, 161–172 (1988).

Melechovský, J., Prins, C. and Calvo, R. W., "A metaheuristic to solve a location-routing problem with non-linear costs", *Journal of Heuristics* 11, 375–391 (2005).

Min, H., Jayaraman, V. and Srivastava, R., "Combined location-routing: a synthesis and future research directions", *European Journal of Operational Research* 108, 1–15 (1998).

Salhi, S. and Rand, G. K., "The effect of ignoring routes when locating depots", *European Journal of Operational Research* 39, 150–156 (1989).

Srivastava, R., "Alternate solution procedures for the location routing problem", *Omega International Journal of Management Science* 21, 497–506 (1993).

Tüzün, D., and Burke, L. I., "A two-phase tabu search approach to the location routing problem", *European Journal of Operational Research* 116, 87–99 (1999).

Wu, T.H., Low, C. and Bai, J.W., "Heuristic solutions to multi-depot location-routing problems", *Computers & Operations Research* 29, 1393–1415 (2002).

HEURISTIC APPROACHES FOR A TSP VARIANT: THE AUTOMATIC METER READING SHORTEST TOUR PROBLEM

Jing Dong, Ning Yang, and Ming Chen
Department of Civil & Environmental Engineering, Glenn L. Martin Hall, University of Maryland, College Park, MD 20742

Abstract: This paper addresses the automatic meter reading shortest tour problem (AMRSTP), a variant of the traveling salesman problem (TSP). The AMRSTP can be formulated as a mixed-integer nonlinear program (MINLP), but solving for the exact solution is impractical. Therefore, two heuristic approaches, a clustering-based algorithm and a convex hull-based algorithm, are proposed to find near-optimal feasible solutions. The algorithms are tested on various datasets, and the numerical results show that both heuristic algorithms perform effectively and efficiently.

Key words: automatic meter reading shortest tour problem; mixed-integer nonlinear program; traveling salesman problem; clustering; convex hull.

1. INTRODUCTION

Automatic meter reading (AMR) was first tested in the early sixties when trials were conducted by AT&T in cooperation with a group of utilities and Westinghouse [1]. Nowadays, AMR has been widely utilized for theft detection, outage management, customer energy management, load management, on/off services, and distributed automation by more and more utility companies since it holds down the cost while increasing the accuracy compared to the traditional labor-intensive meter reading method [2]. By using radio frequency identification (RFID) tags that allow the equipped utility trucks to remotely collect and transmit data, AMR does not require the meter readers to visit each customer's residence, leading to less operating cost and enhanced safety and security.

The AMR systems consist of two parts: RFID tags and a truck-mounted reading device. Each RFID tag, connected to a physical meter, can encode the identification number of the meter and its current reading into digital signals; the truck-mounted reading device can collect the data automatically when it approaches the RFID tags within a certain distance. Given this situation, utility companies would like to design the vehicle routes so as to cover all the customers in the service area and minimize the total tour length or the total cost. The problem is similar to the traveling salesman problem (TSP) except that the tour does not necessarily visit each customer node as long as all the meters can be read from a predefined distance. We call this TSP variant the "automatic meter reading shortest tour problem" (AMRSTP). In this study we assume the AMRSTP is defined in a Euclidean plane.

The AMRSTP is a newly emerging problem; to the best of our knowledge, no previous work on this specific problem has been formally published. In this paper, we formulate this problem as a mixed-integer nonlinear program (MINLP) and use heuristic approaches to solve it.

The paper is organized as follows. In section 2 we propose a mathematical formulation of the AMRSTP. Then we present in section 3 two heuristic approaches, a clustering-based algorithm and a convex hull-based algorithm, to solve the problem approximately. Section 4 provides numerical examples and the results. Finally, conclusions and some future research directions are discussed.

2. FORMULATION

The objective of the AMRSTP is to find an optimal route that minimizes the total distance the utility truck travels, given the locations of the depot, all customers' residences, and the predetermined detecting radius. There is a large body of literature existing on the classic TSP [3, 4]. Previous studies on TSP have assumed that the tour starts from the depot, visits each customer node and then goes back to the depot. In the AMRSTP, however, the tour can cover a customer node without physically visiting this node, provided that the distance between the customer node and the tour is no more than a given radius. Hence for each customer node a "supernode" is introduced to indicate the location that the tour actually visits to cover the customer node. Although the standard TSP is a special case of the AMRSTP, in which the effective radius equals zero, the slight variation makes the AMRSTP significantly more difficult to solve than the TSP, because the locations of supernodes become part of the decision variables in the optimization problem. As one can see, even an exhaustive search might not be applicable since the search region is continuous and there are infinite feasible solutions.

The AMRSTP can be mathematically formulated as a mixed-integer nonlinear program, which is a revised version of the TSP formulation.

2.1 Notation

The following notation is used in the sequel.

n = the number of customer nodes

R = the effective radius

x_i = the x-coordinate of customer node i, for $i = 1,\cdots\cdots,n$

y_i = the y-coordinate of customer node i, for $i = 1,\cdots\cdots,n$

x_0' = the x-coordinate of the depot

y_0' = the y-coordinate of the depot

x_i' = the x-coordinate of supernode i, for $i = 1,\cdots\cdots,n$, which can cover customers i in the sense that the distance between customer node i and supernode i is within effective radius R

y_i' = the y-coordinate of supernode i, for $i = 1,\cdots\cdots,n$

x_{ij} = a binary variable, where x_{ij}= 1 if supernodes (or depot node) i and j are connected by a directed arc from i to j; otherwise x_{ij}= 0, for $0 \le i \le n$ and $0 \le j \le n$.

2.2 Mixed-integer nonlinear program formulation

The problem can be formulated as a mixed-integer nonlinear program (MINLP). Given the customer locations (x_i, y_i), for $i = 1,\cdots\cdots,n$, the depot location (x_0', y_0') and the effective radius R, the problem seeks to find the locations of supernodes (x_i', y_i'), for $i = 1,\cdots\cdots,n$ and the TSP tour x_{ij}, for $0 \le i \le n$ and $0 \le j \le n$ so as to minimize the total tour length while covering all the customer nodes. Note that multiple customer nodes can share the same supernode, i.e., at a single supernode one can collect the data from all residences within the effective radius at the same time. In this case, the problem generates identical supernodes (identical x and y coordinates) for these customer nodes. Then a minimum time/cost tour is searched to visit all the supernodes including the depot. Therefore, the problem is formulated as follows.

$$Min \quad \sum_{i,j} \sqrt{(x'_i - x'_j)^2 + (y'_i - y'_j)^2} \cdot x_{ij}$$

$$\text{s.t. } (x'_i - x_i)^2 + (y'_i - y_i)^2 \leq R^2 \text{ for } i = 1, \cdots\cdots, n; \tag{1}$$

$$\sum_j x_{ij} = 1 \text{ for } i = 0, \cdots\cdots, n; \tag{2}$$

$$\sum_i x_{ij} = 1 \text{ for } j = 0, \cdots\cdots, n; \tag{3}$$

$$\sum_{i \in S} \sum_{j \in S} x_{ij} \leq |S| - 1 \text{ for } S \subset N, \quad 2 \leq |S| \leq n; \tag{4}$$

$$x_{ij} \in \{0,1\}, \text{ for } i = 0, \cdots\cdots, n \text{ and } j = 0, \cdots\cdots, n. \tag{5}$$

$N = \{0,1,2,\cdots\cdots,n\}$, is a set of nodes including all the supernodes and the depot. Constraint (1) ensures that each customer node is covered by its corresponding supernode within the effective radius *R,* while constraints (2) and (3) ensure that each supernode (and depot) is visited exactly once on a tour. Constraint (4) is the usual subtour elimination constraint for a TSP.

3. METHODOLOGY

The mixed-integer nonlinear optimization problem formulated in the previous section involves both integer variables x_{ij}'s and continuous variables (x'_i, y'_i)'s and introduces nonlinearities in the objective function and constraints. Although some research has been done to solve MINLPs, solving a large-scale problem for a global optimum is still time consuming, if not implausible. Therefore heuristic algorithms are proposed in this section to search for near-optimal feasible solutions for the AMRSTP. Our approaches consist of three modules: 1) supernode generation, which determines the supernode set; 2) TSP search module, which solves a traveling salesman problem over the supernode set and the depot; and 3) tour improvement module, which shrinks the tour by taking advantage of radius coverage.

3.1 Supernode generation

The main purpose of the supernode generation module is to find a set of points, or supernodes, which can cover all the customer nodes within the given radius and thus to convert the AMRSTP to a conventional TSP. The general guideline for supernode generation is either to minimize the number of supernodes or to locate the supernodes compactly in an attempt to obtain a shorter tour in subsequent procedures. Accordingly, we employ two different approaches: a clustering-based algorithm and a convex hull-based algorithm.

3.1.1 Clustering-based algorithm

The idea of the clustering-based algorithm is to minimize the number of supernodes by performing distance-based clustering for a given customer node set. A cluster can be viewed as a group of similar objects. The clustering problem has been extensively addressed in the literature in a variety of contexts, and many clustering algorithms have been proposed over the years such as K-means, Fuzzy C-means, Hierarchical clustering and Mixture of Gaussians [5, 6]. A simple version of the K-means algorithm [7] is employed in the AMRSTP.

Specifically, in the clustering-based approach customer nodes are clustered or grouped based on their relative distances, i.e., two or more customer nodes can be clustered only if they are close enough to be covered by a circle with the predetermined radius. The center of this circle is referred to as a supernode. Therefore assigning the customer nodes into as few as possible clusters (circles) is equivalent to finding the minimal number of supernodes. The clustering-based supernode generation module includes three steps. First, generate an initial set of supernodes that can cover the whole plane. Second, remove redundant supernodes that cover no customer node. Third, further reduce the number of supernodes by employing a merging technique.

1. Generate an initial supernode set.

The initial set of supernodes is constructed in the way that the whole Euclidean plane containing all the customer nodes can be covered. Specifically the plane is first tiled with hexagons with edge length R. Then circles are constructed at the center of the hexagons with a radius of R. Thus the plane is completely covered by the circles, with some overlapped area. The initial set of supernodes is defined as the centers of the circles.

After locating the initial set of supernodes, each customer node is assigned to a supernode if the distance between them is at most R, in other words, if the customer node is covered by the supernode. One customer node might be covered by two or three (extreme case) supernodes due to the existence of overlap. Such customer nodes are called overlapped customer nodes, which are tracked to reduce the number of supernodes in the next step. The process is shown in Fig. 1.

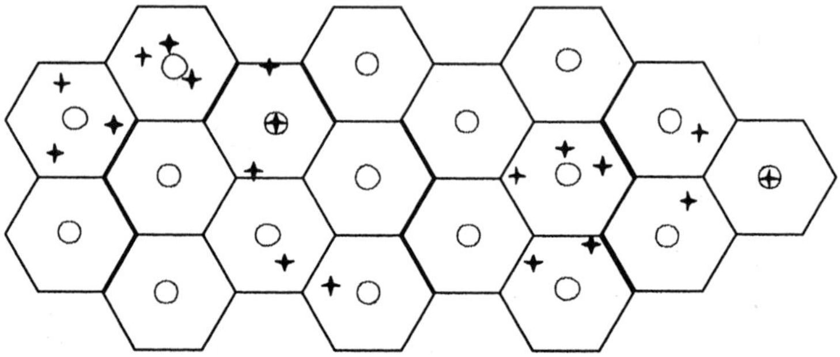

Figure 1. Initial supernode set generation

2. Remove redundant supernodes.

Two types of supernodes are labeled as redundant: the empty supernode that does not cover any customer node; and the overlapping supernode that only covers overlapped customer nodes (for example, the two dashed circles in Fig. 2). In this step each supernode in the initial set is examined and is removed if it is redundant. Deleting the redundant supernodes can maintain the full coverage of all the customer nodes.

One can start checking the overlapping supernodes from the one that covers the least number of overlapped customer nodes to the one that covers the most. By doing this, one might be able to remove more overlapping supernodes. Once an overlapping supernode is removed, the overlapped customer nodes it covers might not be overlapped any longer. The labels need to be updated for these customer nodes accordingly.

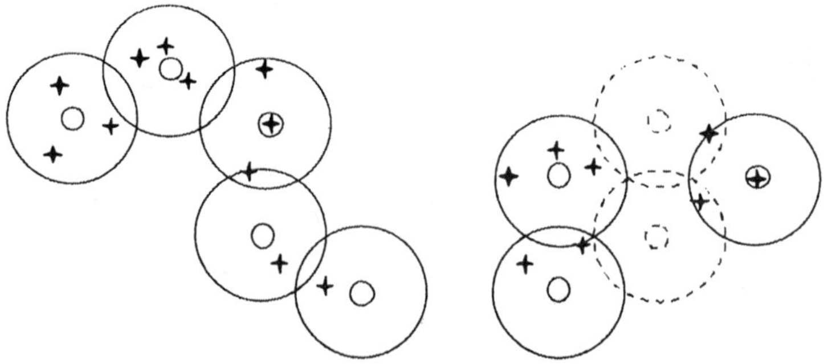

Figure 2. Remove redundant supernodes

3. Merge supernodes.

In this step, the number of supernodes can be further reduced by merging two supernodes (for example, the two dashed circles in Fig. 3) into a new one. First, for each pair of the remaining supernodes, take the minimal and maximal values of x-coordinates and y-coordinates among all the customer nodes that are covered by this pair of supernodes: x_{min}, x_{max}, y_{min} and y_{max}. Then, consider the midpoint of the extreme x's and y's with the x-coordinate and y-coordinate $(\frac{x_{min}+x_{max}}{2}, \frac{y_{min}+y_{max}}{2})$ as a potential new supernode. Finally, check if all the customer nodes that are covered by the original two supernodes can still be covered by the potential new supernode; if yes, replace these two supernodes with the new one. The process stops when no more merging is feasible.

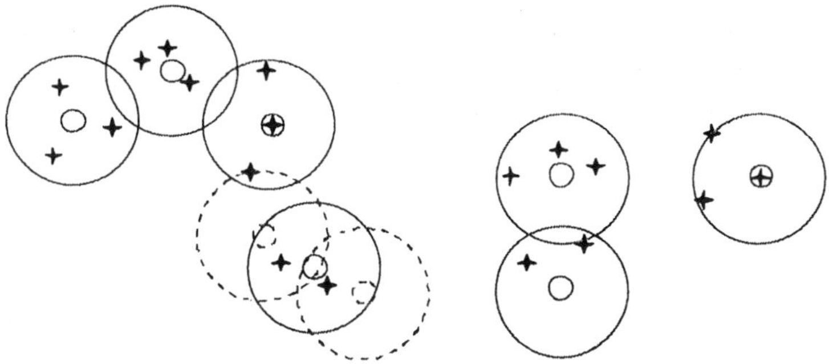

Figure 3. Merge two supernodes

After the above three steps, a set of supernodes is obtained and can be used for the TSP search module.

3.1.2 Convex hull-based algorithm

MacGregor and Ormerod (1996) [8] proposed the convex hull hypothesis, which advocates the use of the convex hull as part of a strategy to create a TSP tour to explain why a high quality TSP solution can be found by trying to obtain a visually attractive solution. Based on the similar idea that a more visually attractive set of supernodes might be found by constructing the convex hulls of customer nodes iteratively, a convex hull-based approach for solving the AMRSTP is developed. The goal of the convex hull procedure is to find a compact set of supernodes that covers all the customer nodes. The set is sought by the following iterative procedure.

Step 0. Initialization. Put all the customer nodes into the "current list" and put the depot into the "supernode list" (see Fig. 4).

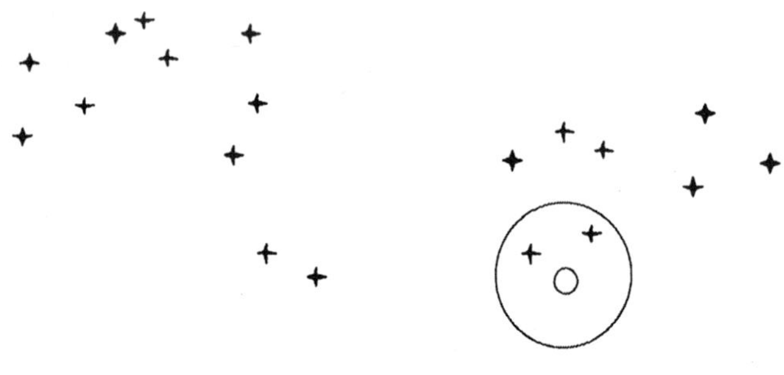

Figure 4. Initialization

Step 1. Scan the "current list" and remove the customer node from the "current list" if it can be covered by at least one supernode in the "supernode list", for example, the two dashed stars in Fig. 4 that can be covered by the depot S are removed from the "current list".

Step 2. Form a convex hull over all the customer nodes in the "current list" and put the vertices (X_i, Y_i), $i = 1, \cdots, m$ into the "convex hull list". The Quickhull algorithm, which was developed by Barber and etc. (1996) [9], is employed for computing the convex hull.

Step 2.0: Find the centroid $O(X_o, Y_o)$ of the vertices in the current "convex hull list", where $X_o = \sum_{i=1}^{m} X_i \Big/ m$ and $Y_o = \sum_{i=1}^{m} Y_i \Big/ m$.

Step 2.1: Select one vertex from the "convex hull list", remove this vertex (named node A) from the list and update $m = m - 1$. If $|AO| > R$, find a point B which is R distant from A along $\overrightarrow{AO}$ and define B as a new supernode; otherwise, define the centroid O as a new supernode. This step is illustrated in Fig. 5.

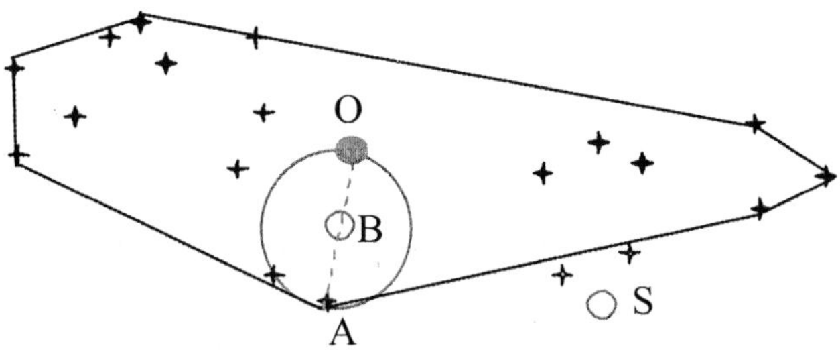

Figure 5. Locate a supernode

Step 2.2: Remove all the customer nodes that can be covered by the new supernode from the "current list".

Step 2.3: Check if the "convex hull list" is empty (*m=0*). If not, go to step 2.0; otherwise, go to step 3.

Step 3: Check if the "current list" is empty. If not (as shown in Fig. 6), go to step 1, otherwise, terminate the procedure and output the "supernode list" (see Fig. 7).

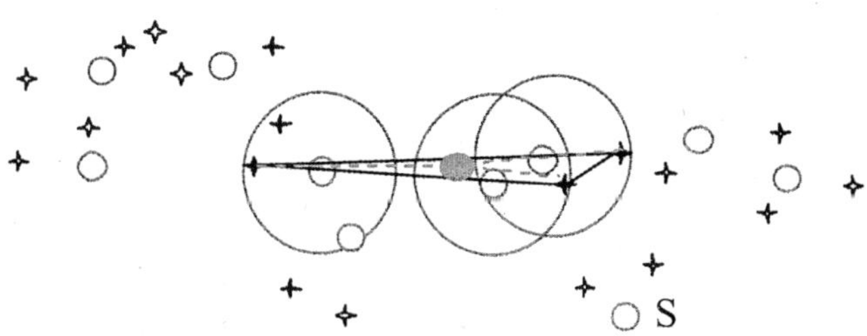

Figure 6. Construct a new convex hull

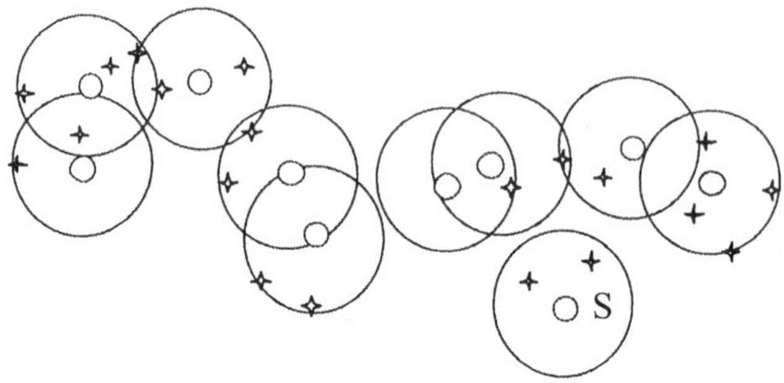

Figure 7. Result of the convex hull-based algorithm

3.2 TSP search

Once a feasible set of supernodes is generated from the previous module, a TSP tour is searched over the existing supernode set and the depot. In the literature a number of algorithmic and heuristic approaches were proposed to find optimal or near-optimal solutions for the TSP. A good review of the TSP heuristic algorithms can be found in Gutin and Punnen (2002) [4]. Among all cases of TSP, the Euclidean TSP is a special one. All methods that have been designed for TSP can be applied to the Euclidean TSP. Besides, special properties hold for the Euclidean TSP [10]. Polynomial-time approximation schemes for the Euclidean TSP have been developed by Arora (1996) [11] and Mitchell (1996) [12]. If H is the convex hull of the nodes in two-dimensional space, the order in which the nodes on the boundary of H appear in the optimal tour will follow the order in which they appear in H [13], which provides the conceptual basis for some intuitive heuristic algorithms [14]. Considering the computational effort and the solution quality, we employ a heuristic algorithm for the Euclidean TSP, including the convex hull insertion algorithm for tour construction and the simulated annealing procedure for tour improvement.

3.2.1 TSP tour construction: convex hull insertion algorithm

Convex hull insertion algorithm was proposed by Stewart and Bodin and it was shown to perform with remarkable speed and surprising accuracy [14]. The procedure is depicted as following. The cost c_{ij} represents the Euclidean distance between node i and node j.

Step 1. Form a convex hull of the set of nodes. The hull gives an initial subtour.

Step 2. For each node k not yet contained in the subtour, decide between which two nodes i and j on the subtour to insert node k. In other words, for each such k, find (i, j) such that $c_{ik} + c_{kj} - c_{ij}$ is minimal.

Step 3. From all (i, j, k) found in step 2, determine the (i^*, j^*, k^*) such that $(c_{i^*k^*} + c_{k^*j^*}) / c_{i^*j^*}$ is minimal.

Step 4. Insert node k^* in subtour between nodes i^* and j^*.

Step 5. Repeat step 2 through 4 until a Hamiltonian cycle is obtained.

3.2.2 TSP tour improvement: simulated annealing algorithm

The simulated annealing (SA) search process adapts a stochastic computational technique to find global or nearly global optimal solutions to combinatorial problems. SA was first developed by Metropolis et al., in 1953 [15], to simulate the annealing process of crystals on a computer and then adapted to solve combinatorial optimization problems by Kirkpatrick et al. [16]. The SA algorithm exploits the analogy between annealing solids and solving combinatorial optimization problems and applies neighborhood search scheme that avoids being trapped in a local extreme by sometimes moving in a locally worse direction. The following SA procedure is implemented for TSP tour improvement [17].

Step 1. Generate an initial tour X using convex hull insertion algorithm. Set the values of the initial temperature T_s, the stop temperature T_f and the temperature decrement ΔT. Choose a replication factor $K_{\max}$. Initialize $T_0 = T_s$, $K = 0$, and $J = 0$.

Step 2 If the stopping criterion ($T_J \le T_f$) is not satisfied, perform the step 2.1 through 2.4.

Step 2.1. Execute one step of the 2-OPT and obtain a modified tour X'. Evaluate the change in the TSP tour length $\Delta C_t = C_t(X') - C_t(X)$. If $\Delta C_t \ge 0$, then go to step 2.2; otherwise, go to step 2.3.

Step 2.2. ($\Delta C_t \ge 0$) Select a random variable $\alpha \in U(0,1)$. If $\alpha < prob(\Delta C_t) = \exp(-\Delta C_t / T_J)$, then go to step 2.3; else if $\alpha > prob(\Delta C_t)$, then reject the modification and go to step 2.1.

Step 2.3. ($\Delta C_t < 0$ or $\alpha < prob(\Delta C_t)$) Accept the 2-OPT exchange and the new value of the tour length $C_t(X')$. Set $X = X'$.

Step 2.4. If $K < K_{\max}$, then set $K = K + 1$; else if $K = K_{\max}$, then $J = J + 1$, $K = 1$, $T_J = T_{J-1} - \Delta T$. Go to step 2.1.

Step 3. Output the TSP tour.

3.3 Tour improvement for the AMRSTP

The key concept of the AMRSTP tour improvement procedure is to shrink the tour obtained from the TSP search module by moving the supernodes in a way that the tour length could be further reduced.

The AMRSTP tour improvement procedure follows three steps.

Step 1. For each supernode i in the current TSP tour, draw a line between its adjacent two supernodes, namely supernodes $(i-1)$ and $(i+1)$, and take the midpoint m. Connect the supernode i with the midpoint m and obtain a directed line segment d that starts from the supernode i and ends at the midpoint m.

Step 2 Move the supernode i along d with a predefined step length. Check if all the customer nodes are still covered. If yes, move the supernode further toward the mid-point m; otherwise, move back to the previous location and go to step 3. Repeat this process until supernode i reaches the mid-point m.

Step 3. Repeat step 1 and 2 until all the supernodes are scanned and an improved tour is obtained.

4. NUMERICAL EXAMPLES

To evaluate the effectiveness and efficiency of the algorithms proposed in the previous section, numerical tests were conducted. Factors that might affect the performance of the proposed algorithms include the effective radius, number of customer nodes located within a certain area, distribution of these customer nodes, etc. The numerical experiments primarily focused on the first two factors, i.e., radius and number of customer nodes.

4.1 Experimental design

The datasets were generated with number of customer nodes varying from 100 to 1000 in increments of 50. For the datasets with 300 customer nodes, 10 different radii were examined varying from 2 to 20 in increments of 2. Overall, 190 cases were constructed and tested, each of which involved one radius and a particular number of customer nodes. For comparison purposes, all the customer nodes were generated randomly within a 100×100 predefined study area. Five replicates of data sets were generated for each case and the results were averaged over the 5 outputs.

Both the resulting tour lengths and running times were tracked for assessing the proposed algorithms. Three scenarios were compared: solving the AMRSTP using the clustering-based algorithm, solving the AMRSTP using the convex hull-based algorithm, and solving TSP over all the customer nodes. The TSP scenario was introduced to demonstrate the advantage of adopting an automatic meter reading method as well as to provide an upper bound on the solution to the AMRSTP. In order to obtain a tighter upper bound, the TSP was solved by the Concorde's TSP solver [18], which has been used to obtain the optimal solutions to 107 of the 110 TSPLIB (a library of sample instances for the TSP and related problems from various sources and of various types) instances [19]. At the expense of running time the TSP solutions obtained from Concorde solver are either optimal or very close to optimal solutions.

Algorithms were coded in C++ programming language and tested on Pentium 4 computer with 3.20GHz CPU and 2GB of RAM.

4.2 Numerical results

Figures 8 and 9 show the resulting tours generated by employing the clustering-based algorithm and the convex hull-based algorithm, respectively. Figure 10 shows the corresponding TSP tour for the same problem that involves 300 customer nodes and the effective radius of 10. As expected, allowing the data collector to use the remote sensor can significantly reduce the total tour length (or cost) as compared to the corresponding TSP solution. The extent to which the tour length can be reduced depends on the radius as well as the amount and the distribution of the customer nodes. In this particular example, the TSP tour length can be reduced by 54.05% if the clustering-based algorithm is employed or 58.23% if the convex hull-based algorithm is employed.

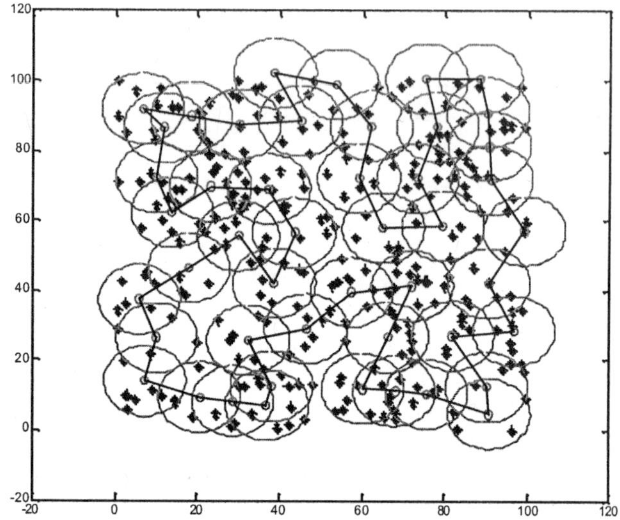

Figure 8. Tour generated from the clustering-based algorithm

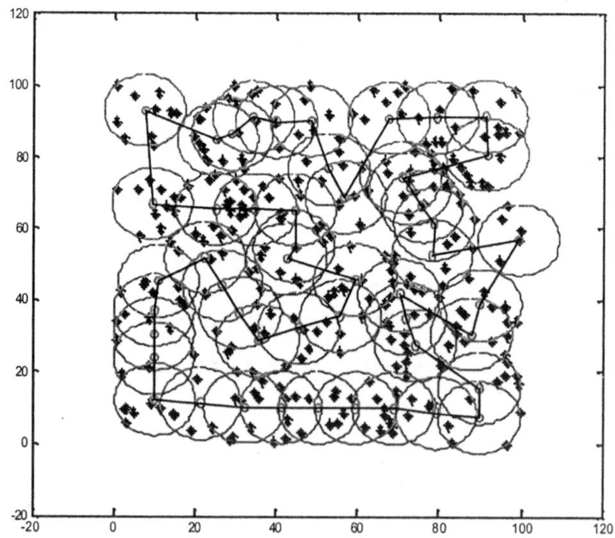

Figure 9. Tour generated from the convex hull-based algorithm

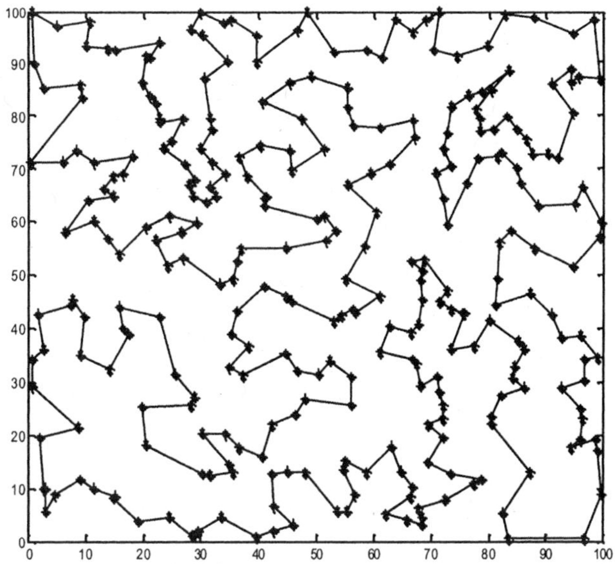

Figure 10. TSP tour

Figure 11 shows the resulting tour lengths (averaged over 5 duplicates) with the number of customer nodes varying from 100 to 1000 in increments of 50 and fixed radius of 10. Tour lengths generated by employing the two heuristic algorithms and the TSP are shown in the same figure with different legends. As shown in Fig. 11, the total tour lengths vary with different numbers of nodes. In general, more customer nodes result in a longer tour length, as one might anticipate. The tour length remained around a "fixed" value (600 for the convex hull-based algorithm and 650 for the clustering-based algorithm) when the number of customer nodes exceeds a certain value, i.e., 650 in this example. The rationale is that once the tour covers the whole study area, introducing additional customer nodes no longer significantly affects the tour length. Compared with the TSP solution, both algorithms can considerably reduce the tour length, especially when the problem size gets larger. For example, the TSP tour length can be reduced by approximately 40% in the examples with 100 customer nodes and by approximately 70% in the examples with 1000 customer nodes. The results indicate that the introduction of an automatic meter reading technique can bring more benefits to an area with larger residential density. Moreover, Figure 11 shows that the convex hull-based algorithm outperforms the clustering-based algorithm in terms of the total tour length, which is achieved at the expense of running time as illustrated in Fig. 12.

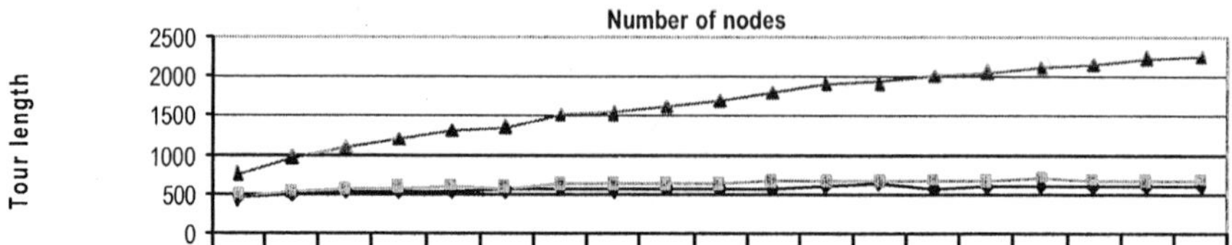

	100	150	200	250	300	350	400	450	500	550	600	650	700	750	800	850	900	950	1000
Convex Hull	454	492	525	509	527	546	570	554	575	570	572	602	617	574	608	589	594	606	595
Clustering	474	527	542	572	601	579	618	633	637	636	661	651	657	676	662	694	686	661	674
TSP	759	951	1083	1190	1299	1363	1515	1545	1609	1700	1803	1891	1933	2012	2053	2115	2174	2234	2272

Convex Hull — Clustering — TSP

Figure 11. Total tour lengths with varying numbers of customer nodes

Figure 12 demonstrates that the running time depends on the problem size, although for the tested examples running times are less than 1 second when employing both heuristic algorithms. However, solving the TSP by the Concorde's solver takes several minutes to an hour in the tested datasets. The running times for solving the TSP are not provided, since they are not comparable to the AMRSTP heuristic algorithms. The result implies that the proposed heuristic algorithms might be applied to a large-scale problem with reasonable running time.

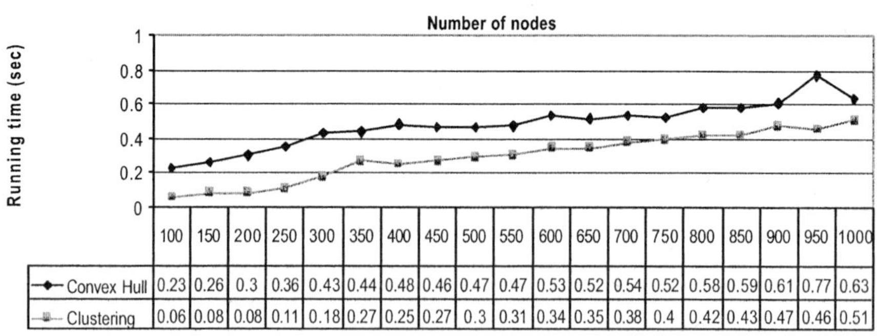

	100	150	200	250	300	350	400	450	500	550	600	650	700	750	800	850	900	950	1000
Convex Hull	0.23	0.26	0.3	0.36	0.43	0.44	0.48	0.46	0.47	0.47	0.53	0.52	0.54	0.52	0.58	0.59	0.61	0.77	0.63
Clustering	0.06	0.08	0.08	0.11	0.18	0.27	0.25	0.27	0.3	0.31	0.34	0.35	0.38	0.4	0.42	0.43	0.47	0.46	0.51

Convex Hull — Clustering

Figure 12. Running times comparison

The performance of both algorithms with varying radii is also evaluated. The tested examples include 300 customer nodes. Figure 13 shows the tour length (averaged over 5 replicates) obtained by employing these two algorithms. The radii examined ranges from 2 to 20 in increments of 2. Intuitively, a larger radius results in a shorter tour length, as is confirmed by the results. In this particular example, the clustering-based algorithm outperforms the convex hull-based algorithm when the radius is small (less than 6 in the example), while the convex hull-based algorithm performs better when the radius gets larger (exceeds 6 in the example). Similar results were found in other tested examples with different numbers of customer nodes. These initial findings are interesting.

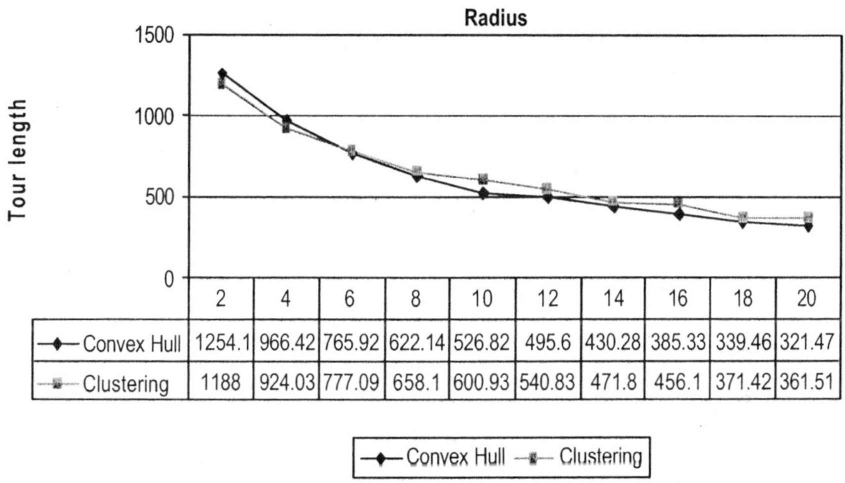

Radius	2	4	6	8	10	12	14	16	18	20
Convex Hull	1254.1	966.42	765.92	622.14	526.82	495.6	430.28	385.33	339.46	321.47
Clustering	1188	924.03	777.09	658.1	600.93	540.83	471.8	456.1	371.42	361.51

Figure 13. Total tour lengths vary with radius

In addition, all the tour lengths listed here from two heuristic algorithms are the final tour lengths after performing three procedures: supernode generation, TSP search and tour improvement. Numerical experiments were also employed to test the benefit of the tour improvement procedure. The results show that for both heuristic algorithms the tour lengths can be significantly decreased after the tour improvement step. Specifically, in the tested datasets, the average improvements are approximately 20% for the clustering-based algorithm and 5% for the convex hull-based algorithm. It shows that the convex hull-based algorithm could achieve relatively better initial solutions after performing the first two procedures.

5. CONCLUDING REMARKS

In this paper, a new research problem, the automatic meter reading shortest tour problem, is introduced and formulated as a mixed-integer nonlinear program. Given the complexity of the problem, a clustering-based algorithm and a convex hull-based algorithm have been developed to search for near-optimal solutions. Experimental results show that both algorithms are computationally efficient while providing good quality solutions. The performance of the proposed algorithms depends on the number and the distribution of customer nodes as well as the effective radius. While more systematic testing is necessary to draw a strong conclusion, it was found that the clustering-based algorithm performed better for problems with small radii, while the convex hull-based algorithm performed better for problems with large radii

Two topics might be of great interest in future research work. First, the AMRSTP was considered in a Euclidean plane in which people have complete freedom to choose a route. However, the real street network eliminates many options by providing linkages between only certain pairs of locations. The constraint reduces the number of feasible solutions, but it also invalidates the proposed algorithms. New or modified algorithms need to be explored to solve the generic AMRSTP with network constraints. Furthermore, in this study a single vehicle serves the entire study area. In reality, multiple vehicles might be involved and the work load may need to be split. As an extension, the AMRSTP for multiple vehicles can be examined, taking into consideration the work load balance among all vehicles.

6. ACKNOWLEDGEMENTS

The authors wish to express their sincere gratitude to Dr. Bruce Golden for providing us with the exciting research topic and offering his most valuable help.

REFERENCES

1. T. D. Tamarkin (1992), Automatic Meter Reading, *Public Power*, 50(5).
2. Carlos A. Osorio Urzua (2004), Bits of Power: The Involvement of Municipal Electric Utilities in Broadbank Services, master thesis, MIT.
3. E.L. Lawler, J.K. Lenstra, A.H.G. Rinnooy, D.B. Shmoys (1985), *The Traveling Salesman Problem: A Guided Tour of Combinatorial Optimization*, Wiley, New York.

4. G. Gutin and A.P. Punnen (2002), *Traveling Salesman Problem and Its Variations*, Kluwer Academic Publishers.
5. A. K. Jain, M. N. Murty, and P. J. Flynn (1999), Data clustering: a review, *ACM Computing Surveys*, 31(3): 264-323.
6. C. Fraley and A. E. Raftery (1998), How many clusters? which clustering method? answers via model-based cluster analysis, *The Computer Journal*, 41(8):578-588.
7. J. B. MacQueen (1967), Some Methods for classification and Analysis of Multivariate Observations, *Proceedings of 5-th Berkeley Symposium on Mathematical Statistics and Probability*, Berkeley, University of California Press, 1:281-297.
8. J. N.MacGregor, T. C. ,Ormerod (1996), Human performance on the traveling salesman problem, *Perception & Psychophysics*, 58: 527-539.
9. C. B. Barber, D. P. Dobkin, ,and H.T.Huhdanpaa (1996), The Quickhull algorithm for convex hulls, *ACM Trans. on Mathematical Software*, 22(4): 469-483.
10. R. C. Larson and A. R. Odoni Eds. (1981), *Urban Operations Research*, Prentice-Hall, NJ.
11. S. Arora (1996), Polynomial time approximations schemes for Euclidean TSP and other geometric problems, *Proc. 37th IEEE Foundations of Computer Science*: 1-10.
12. J. Mitchell (1996), Guillotine subdivisions approximate polygonal subdivisions: Part II - A simple polynomial-time approximation scheme for geometric k-MST, TSP, and related problems, University at Stony Brook, Part I appears in *SODA'96*, 402-408.
13. S. Eilon, C. Watson-Gandy and N. Christofides (1971), *Distribution Management*, Griffin, London.
14. B. Golden; L. Bodin; T. Doyle; W. Stewart Jr.(1980), Approximate Traveling Salesman Algorithms, *Operations Research*, 28(3): 694-711.
15. N. Metropolis, A. W Rosenbluth., M. N. Rosenbluth,, and A. H. Teller (1953), Equation of State Calculation by Fast Computing Machines, *Journal of Chemical Physics*, 21: 1087-1092.
16. S.Kirkpatrick , C. D. Gelatt,, and M. P. Vecchi (1983), Optimization by Simulated Annealing *Science*, 220: 671-680.
17. Christopher C. Skiscim, Bruce L. Golden, Optimization by Simulated Annealing: A Preliminary Computational Study for the TSP, *Proceedings of the 1983 Winter Simulation Conference*: 523-535.
18. Concorde TSP solver; http://www.tsp.gatech.edu/concorde/index.html
19. TSPLIB; http://www.iwr.uni-heidelberg.de/groups/comopt/software/TSPLIB95/

THE GENERALIZED TRAVELING SALESMAN PROBLEM: A NEW GENETIC ALGORITHM APPROACH

John Silberholz[1] and Bruce Golden[1]
[1]R.H. Smith School of Business, University of Maryland, College Park, MD 20742

Abstract: The Generalized Traveling Salesman Problem (GTSP) is a modification of the Traveling Salesman Problem in which nodes are partitioned into clusters and exactly one node from each cluster is visited in a cycle. It has numerous applications, including airplane routing, computer file sequencing, and postal delivery. To produce solutions to this problem, a genetic algorithm (GA) heuristic mimicking natural selection was coded with several new features including isolated initial populations and a new reproduction mechanism. During modeling runs, the proposed GA outperformed other published heuristics in terms of solution quality while maintaining comparable runtimes.

Key words: Generalized traveling salesman problem; genetic algorithm; combinatorial optimization

1. INTRODUCTION

The Generalized Traveling Salesman Problem (GTSP) is a variant of the well-known Traveling Salesman Problem (TSP). As in the TSP, the graph considered consists of n nodes, and the cost between any two nodes is known. The GTSP differs from the TSP in that the node set is partitioned into m clusters. An optimal GTSP solution is a cycle of minimal cost that visits exactly one node from each cluster.

The GTSP has numerous real-world applications including welfare agency routing, in which agencies have specializations and a dient needs only to visit one agency with each specialization, desiring to minimize travel costs, as described in [13]. Other applications include those in airplane routing [9], mail delivery [6], warehouse order picking [9], material flow system design [6], vehicle routing [6], and computer file sequencing [5].

Finding efficient solutions to complex GTSP problems is vital to many disciplines, especially as agencies struggle to cope with today's increased transportation costs due to higher fuel prices.

Several GTSP variations have emerged based upon the specifics of the set of nodes considered. This paper assumes symmetric costs or distances, that is, $c_{ij} = c_{ji}$, where c_{ij} is the cost or distance between nodes i and j. This means that the direction of travel between two nodes doesn't affect the cost. Additionally, some versions of the GTSP require that at least one node from each cluster be visited, instead of exactly one. While these two variations are equivalent as long as the triangle inequality holds, it may cost less to visit extra nodes if the triangle inequality does not hold. This paper assumes that exactly one node from each cluster is visited, an approach that is sometimes called the Equality GTSP (E-GTSP) [2].

Ideally, an exact algorithm, or one that always produces optimal solutions, would be most desirable. However, use of such procedures, like the one presented in [3], is not always feasible, because they tend to have prohibitively long runtimes for problems defined on a large number of nodes or clusters. For instance, the authors of [3] did not attempt to run their exact algorithm on problems larger than 442 nodes or 89 clusters because runtime of their algorithm was rapidly approaching one day. This shortcoming introduces the need for quicker heuristic methods, or approximate algorithms, which provide reasonable solutions to a problem in shorter runtimes. Examples of some heuristics for the GTSP include Snyder and Daskin's Genetic Algorithm (S+D GA) solution [14], Renauld and Boctor's GI^3 heuristic [11], Noon's generalized nearest neighbor heuristic with GI^3 improvement (NN) [11], and Fischetti et al.'s Lagrangian and root-node heuristics [3].

A genetic algorithm (GA) is a heuristic that mimics the process of natural selection. In such an algorithm, a population slowly converges to a final individual with an associated objective value after a number of iterations each of which corresponds to a new generation of that population. To facilitate this, the most desirable solutions within the population are assigned the highest survival rate from one generation to the next.

GAs store a population of chromosomes, each of which is a candidate solution for its corresponding problem (in this case, the GTSP). In each generation (iteration) of the heuristic, several operations are performed on the chromosomes to improve the overall fitness (i.e., cost) of the population. First, replication can occur, in which chromosomes are directly passed along to the next generation. These chromosomes are selected with a weighting system favoring better (lower) total cycle costs. Then, crossover, or reproduction, can occur — the GA equivalent of two parents mating and producing two children, both of whom bear a resemblance to each parent. Crossover operators that facilitate this reproduction include the partially mapped crossover (PMX) found in [4], the maximal preservative crossover

(MPX) found in [8], and the ordered crossover (OX) found in [2]. A comparison of different crossovers used for the TSP can be found in [15]. Finally, mutation, a process that alters randomly selected portions of the chromosome, is also possible. For the GTSP, a common method of mutation is inversion, following [7], which is considered later in this paper.

Using the basic structure of a GA as defined in [7], this paper explores effective alternative genetic structures and crossover operators. This paper supplements the current literature by testing an effective algorithm that uses these GA improvements. The proposed GA generates high quality solutions to instances of the GTSP in reasonable runtimes.

2. THE GENETIC ALGORITHM

Data were collected on a Dell Dimension 8400 with 1.0 GB RAM and a 3.0 GHz Intel Pentium 4 processor, using programs coded in Java 1.4 and run on the Eclipse platform. This paper's GA was developed based upon a general discussion of heuristics developed for the TSP in [7]. Due to the simplicity and effectiveness of using a path representation of a TSP, as described in [7], a path representation was used for the storage of GTSP candidate solutions in chromosomes.

2.1 Path representation

In the path representation, the most natural and simplistic way to view GTSP pathways, each consecutive node in the representation is listed in order. For instance, the chromosome (1 5 2) represents the cycle visiting node 1, then node 5, then node 2, and finally returning to node 1. Advantages of this representation include simplicity in fitness evaluation, as the total cost of a cycle can easily be calculated by summing the costs of each pair of adjacent nodes, and the usefulness of the final representation, as it directly lists all of the nodes and the order in which they are visited. However, a shortcoming of this representation is that it carries no guarantee that a randomly selected representation will be valid for the GTSP, because there is no guarantee that each cluster is represented exactly once in the pathway without specialized procedures or repair algorithms.

2.2 Population initialization

At the beginning of the GA, each new chromosome was generated by continuously selecting random nodes and adding them to the new chromosome one by one provided that another node from the same cluster had not already been incorporated. An initial population consists of 50 of these chromosomes, a size which was deemed reasonable considering examples provided in [7].

2.3 Crossover

A novel reproductive method based upon the TSP ordered crossover (OX) operator proposed by Davis in [2] was used. The TSP's OX crossover randomly selects two cut points on one of two parent chromosomes. The nodes between these two points on the first parent are maintained in their same locations, and the remaining non-duplicate nodes from the second parent are placed, in order, at the remaining locations of the offspring, yielding a child containing ordered genetic material from both parents. For instance, from two parents $p_1 = (1 \mid 5\ 4 \mid 3\ 2)$ and $p_2 = (2 \mid 3\ 5 \mid 1\ 4)$, with cut points denoted by vertical bars, the material (genes) between the cut points in p_1, nodes 5 and 4, are maintained and the non-duplicate nodes from p_2, copied in order from after the second breakpoint, are nodes 1, 2, and 3. Insertion of these nodes into the offspring would yield a final chromosome $c_1 = (3\ 5\ 4\ 1\ 2)$. Note that the inserted material, which was added after the second cut point, wraps around to the beginning of the chromosome when it reaches the end, providing for a complete offspring. Maintaining the same cut points, the other offspring would be (4 3 5 2 1). An illustration of the OX operation is provided in [7] on pp. 217-218.

A simple modification to convert this crossover to the GTSP involves insertion of nodes from the second parent whose clusters do not coincide with those of the selected nodes from the first parent.

The initial crossover mechanism was further modified by adding a rotational component. Nodes selected for insertion from the second chromosome were rotated, and numerous orientations of the nodes to be inserted were considered. For instance, instead of simply inserting the nodes from the second parent in the previous example in the order 1-2-3, the orderings 2-3-1 and 3-1-2 were also considered, and the ordering which created an offspring with the least cost was added. Though a large number of orderings are considered for larger subtours from the second parent, little computation time is expended, as only two cost evaluations are needed to determine the effectiveness of a rotation, each directly at a cut point. An additional component of this rotational crossover, which allows reversals of strings to be inserted, was also implemented. This would have yielded the additional consideration of orderings 3-2-1, 2-1-3, and 1-3-2. This reversed insertion is applicable only to a symmetric GTSP, because each reversed string would have to be completely reevaluated for an asymmetric GTSP dataset. This modified crossover, including both the rotational and reverse rotational components, will be referred to as the rotational ordered crossover, or rOX.

Table 1. Example's explicit, symmetric distance matrix

Cluster	1		2		3		4		5		6	
Node	1	2	3	4	5	6	7	8	9	10	11	12
1	0	41	31	86	25	57	7	13	21	19	41	47
2	41	0	38	74	43	98	35	31	11	48	24	69
3	31	38	0	50	89	7	30	74	69	16	20	58
4	86	74	50	0	89	92	34	9	69	13	44	79
5	25	43	89	89	0	56	28	35	68	86	82	83
6	57	98	7	92	56	0	85	52	32	77	31	46
7	7	35	30	34	28	85	0	59	47	36	42	18
8	13	31	74	9	35	52	59	0	43	86	81	74
9	21	11	69	69	68	32	47	43	0	50	16	95
10	19	48	16	13	86	77	36	86	50	0	29	8
11	41	24	20	44	82	31	42	81	16	29	0	19
12	47	69	58	79	83	46	18	74	95	8	19	0

The rOX was further modified with an additional rotational component at the cut points. This operator rotates both of the bordering nodes from the second parent through each of the possible nodes within its cluster, selecting the one that would minimize the final cost of the tour. While this modification required significantly more runtime, it produced better solutions that tended to increase population diversity. It did so by increasing the one-generation survival probability of a promising new orientation of solutions that has not yet been locally optimized but may eventually produce better results than the current best result. This further improvement on the rOX will, hereafter, be referred to as the modified rotational ordered crossover, or mrOX. As this crossover is a defining characteristic of this paper's heuristic, the algorithm presented in this paper will be referred to as the mrOX GA. An example is provided in Table 1. Consider two parents, p_1 and p_2. They are defined based on the distance matrix provided in Table 1, and cut points were selected around the middle two nodes.

$$p_1 = (\ 12 \quad 1 \quad | \quad 3 \quad 10 \quad | \quad 6 \quad 8 \ ) \qquad \text{with cost} = 297$$
$$p_2 = (\ 2 \quad 4 \quad | \quad 6 \quad 8 \quad | \quad 10 \quad 12 \ ) \qquad \text{with cost} = 381$$

Nodes 3 and 10, which are between the cut points in p_1, are in clusters 2 and 5. After the right cut point (with wrap-around), p_2 visits clusters 5, 6, 1, 2, 3, and 4. Removing clusters 2 and 5 (to ensure that chromosome produced contains exactly one constituent of each cluster, making it legal) leaves, in order, clusters 6, 1, 3, and 4. Forward rotation yields the following orderings of clusters —

Table 2. Chromosomes considered in example mrOX crossover

Rotational Crossover							Reverse Rotational Crossover						
pos1	pos2	pos3	pos4	pos5	pos6	cost	pos1	pos2	pos3	pos4	pos5	pos6	cost
6	8	3	10	12	2	317	2	12	3	10	8	6	379
6	8	3	10	11	2	293	2	12	3	10	7	6	362
6	7	3	10	12	2	306	2	11	3	10	8	6	296
6	7	3	10	11	2	282	2	11	3	10	7	6	279
8	12	3	10	2	6	346	12	8	3	10	6	2	408
8	12	3	10	1	6	276	12	8	3	10	5	2	362
8	11	3	10	2	6	315	12	7	3	10	6	2	308
8	11	3	10	1	6	245	12	7	3	10	5	2	262
12	2	3	10	6	8	326	8	6	3	10	2	12	266
12	2	3	10	5	8	318	**8**	**6**	**3**	**10**	**1**	**12**	**215**
12	1	3	10	6	8	297	8	5	3	10	2	12	331
12	1	3	10	5	8	289	8	5	3	10	1	12	280
2	6	3	10	8	12	350	6	2	3	10	12	8	286
2	6	3	10	7	12	244	6	2	3	10	11	8	314
2	5	3	10	8	12	377	6	1	3	10	12	8	238
2	5	3	10	7	12	271	6	1	3	10	11	8	266

6, 1, 3, 4
1, 3, 4, 6
3, 4, 6, 1
4, 6, 1, 3.

Reverse rotation yields the following orderings of clusters —

4, 3, 1, 6
3, 1, 6, 4
1, 6, 4, 3
6, 4, 3, 1.

If the rOX were being performed, the nodes from p_2 in the clusters listed above would be inserted in order to the right of the nodes retained from p_1, wrapping around the chromosome if necessary. However, as an mrOX is being performed, full rotation is completed on the two clusters that border the retained nodes (the first and last clusters listed above). Thus, for the first list of clusters (6, 1, 3, 4), it is clear that the nodes to be inserted from p_2 are 12, 2, 6, and 8. However, in the mrOX, rotating through the bordering clusters also yields orderings 11, 2, 6, 8; 12, 2, 6, 7; and 11, 2, 6, 7. These four possible insertion orders are the top four orderings considered in Table 2. Table 2 contains all of the 32 possible orderings considered by the mrOX crossover, along with the associated cost of each considered pathway. It should be noted that, given a node set with *n* nodes,

m clusters, a distance between cut points of d, and an ordering of clusters from p_2 named w, the number of chromosomes considered is $R * \sum_{i=1}^{m} \beta_i r_i (r_f + r_g)$. R is the reversal constant which equals 2 if reversals are considered (the space is symmetric) and 1 if not, and β_i is the cluster exclusion constant which equals 1 if a cluster is outside of the cut points retained from p_1 and 0 if it is contained between the cut points. $f = o_{(x_i+1) \bmod (m-d)}$ and $g = o_{(x_i-1) \bmod (m-d)}$. r_i is a function that returns the number of nodes in a cluster i, x_i is a function that returns the position of a cluster i in w, and o_q is the cluster at a certain position q in w. This equation returns 32 when considering the example crossover provided in Table 2.

As (8 6 3 10 1 12) is the possible offspring with the bwest cost, 215, this bolded entry in Table 2 becomes the actual offspring of p_1 and p_2. It should be noted that the standard OX crossover, when applied to this situation, returns the chromosome (6 8 3 10 12 2), with cost of 317.

To improve the speed of crossover execution, the distance between cut points on the first parent was increased, decreasing the number of necessary comparisons. The first cut point was randomly selected, and if it was on the right side of the chromosome, the other point was inserted at position $\left\lfloor rand^2 \frac{m}{2} \right\rfloor + 1$. Otherwise, the point was inserted at position $m - \left\lfloor rand^2 \frac{m}{2} \right\rfloor$. In these expressions, *rand* is a random real number on [0, 1) and m is the number of clusters in the dataset.

2.4 Population structure

Additional improvements were made to the fundamental structure of a GA. First, to maintain diversity, no duplicate chromosomes (including rotations or reversals of the same chromosome) were allowed to coexist in a population. This is easily facilitated by maintaining the position of the cluster 1 gene in each of the chromosomes in the population for easier comparison to determine similarity.

Instead of a standard GA structure, which involves the evolution of one population of chromosomes into a final solution, the new structure involves isolating several groups of chromosomes for a relatively short time at the beginning of the solution procedure and using less computationally intensive genetic procedures and local improvement to rapidly generate reasonable solutions. Then, the best chromosomes from each of the smaller populations are merged into a final population, which is improved with a standard genetic algorithm structure. For the algorithm presented in this paper, seven isolated populations were maintained, each containing 50 chromosomes. After none of the populations produced a new best solution in 10 generations, the best 50 solutions from the combined pool of 350 became the final population to be improved. To ensure the speed of convergence of the initial populations, each used the rOX crossover and quicker local improvement heuristics (see Section 2.6).

In each generation, 20 of the 50 chromosomes in the population remained unaltered through replication from the previous generation. Instead of directly selecting these individuals, the thirty non-replicated chromosomes were selected through a spinner procedure, in which each chromosome was given a probability of death (with all probabilities adding to 1), and a spinner was spun to determine which chromosomes died. The affinity for death, ad_i, was calculated as $ad_i = (c_i - c_{best})^{deathPow}$ for each chromosome of index i, where c_i is the cost of that solution, c_{best} is the cost of the best (least cost) solution in the population, and *deathPow* is a constant that controls algorithmic convergence. The *deathPow* was set at 0.375, which was determined by experimentation to provide reasonable population diversities and convergence speeds. The probability of death of each individual chromosome was calculated by dividing each ad_i by $\sum_{i=1}^{50} ad_i$.

2.5 Reproduction

In each generation, the last 30 chromosomes added were individuals produced through reproduction. Parents were determined through a spinner selection similar to that used to determine death. The affinity for reproduction, ar_i, was calculated as $ar_i = (c_{worst} - c_i)^{reprodPow}$ for each chromosome of index i, where c_i is the cost of that solution, c_{worst} is the cost of the worst (most costly) solution in the population, and *reprodPow* is a constant that controls algorithmic convergence. The *reprodPow* was set at 0.375, which was determined by experimentation to provide reasonable population diversities and convergence speeds. The probability of reproduction of each individual chromosome was calculated by dividing each ar_i by $\sum_{i=1}^{50} ar_i$. Individual chromosomes can be selected more than once for reproduction.

Once a list of 30 parents was generated, each pair produced two children. Before isolated populations merged, each child was generated with the rOX crossover, but subsequent generations of offspring were created using the mrOX crossover.

2.6 Local improvement heuristics

Local improvement heuristics, which apply a set of transformations to a single solution, significantly improve the performance of GAs [14]. Thus, the popular *2-opt* local improvement heuristic was implemented. The 2-opt replaces a pair of edges with two edges that shorten the overall path length, if such a pair exists in the chromosome.

Additionally, the swap operator described in [14] was used to further strengthen local optimization in the solution. The swap operator removes each node from the tour and replaces it in every other possible position, selecting the first position that improves overall solution quality. In replacement, the node can be rotated through its cluster. Consider the example below, which uses the distance matrix from Table 1.

The chromosome considered is (2 12 3 5 7 9), with a cost of 302. The first node to be considered is 2. Insertion into each other possible position in the chromosome yields possible solutions (12 2 3 5 7 9), (12 3 2 5 7 9), (12 3 5 2 7 9), (12 3 5 7 2 9), and (12 3 5 7 9 2). The costs of these solutions are, respectively, 366, 309, 367, 316, and 302. Since none of these new positionings produced an improvement in solution cost, the other node in 2's cluster, 1, is considered in each position as a replacement for 2. The first chromosome considered, (1 12 3 5 7 9), has a cost of 290, which is lower than the cost of the initial chromosome considered, and thus becomes the final solution produced by the swap operation.

For the initial isolated populations, a lower level of local optimization was used to shorten runtime, in which the best chromosome found in the previous generation replaces the first chromosome in the current population if it is not already present, and the best chromosome in the current generation receives exactly one two-opt (or one swap if all available 2-opts are exhausted).

After the isolated populations are merged, each child produced with a better fitness (lower cost) than its parents receives full local improvement, which involves carrying out 2-opts until none are available and then swaps until none are available. Since a swap could cause a two-opt to become available, and vice versa, the cycle is repeated until no more local improvements are available. Full local optimization is also used on a randomly selected 5% of the new chromosomes produced through reproduction to improve diversity and solution quality at the cost of increased runtime.

2.7 Mutation

To facilitate mutation and thus improve population diversity, each chromosome in the population had a 5% probability of being selected for mutation, a rate similar to those used in [7]. If selected, two cut points were randomly selected from each chromosome's interior, and the nodes between these two points were reversed. If $p_1 = (1 \mid 5\ 4 \mid 3\ 2)$, with the selected cut points denoted by the vertical bars, then the inverted chromosome $p_1' = (1\ 4\ 5\ 3\ 2)$.

2.8 Termination conditions

The algorithm terminated after the merged population did not produce a better solution for 150 generations. This termination generation count is larger than that of most genetic algorithms because the heuristic proposed has less local optimization than most other approaches.

3. COMPUTATIONAL EXPERIMENTS

The Snyder and Daskin GA (S+D GA) was selected for machine-independent comparison with this paper's mrOX GA both because it is also a genetic algorithm, and thus comparable, and because it produced some of the best heuristic results for the GTSP to date, as detailed in [14]. We implemented the S+D GA, whose attributes are detailed in [14]. In particular, we coded it in Java to produce comparable runtimes and to allow comparisons with our GA for larger datasets than those tested in [14]. The Java implementation had nearly identical performance to the Snyder and Daskin program over the datasets cited in [14], which ranged in size from 48 to 442 nodes. A two-sided paired t-test comparing results of five trials for each dataset considered in [14] with a null hypothesis that the algorithms were identical yielded a p-value of 0.9965, suggesting near-identical results. Because all heuristics rely heavily on random numbers, it is expected that the results are slightly different from the published values.

The datasets tested, as with all testing sets considered in this paper, were acquired from Reinelt's TSPLib [10]. This data source was selected because of easy Internet accessibility at softlib.rice.net, and because most papers concerning GTSP heuristics have used these datasets.

Each dataset was clustered using the procedure "CLUSTERING" described in Section 6 of [3] and implemented in, for example, [11] and [14]. This method clusters nodes based on proximity to each other, iteratively

Table 3. Comparison of heuristic solution qualities and runtimes

Dataset Name	mrOX GA		S+D GA		GI^3		NN		FST-Lagr		FST-Root		B&C
	Pct.	Time	Pct.	Time	Pct.	Time	Pct.	Time	Pct.	Time	Pct.	Time	Time
10ATT48	**0.00**	0.36	**0.00**	0.18	*		*		**0.00**	0.90	**0.00**	2.10	2.10
10GR48	**0.00**	0.32	**0.00**	0.08	*		*		**0.00**	0.50	**0.00**	1.90	1.90
10HK48	**0.00**	0.31	**0.00**	0.08	*		*		**0.00**	1.10	**0.00**	3.80	3.80
11EIL51	**0.00**	0.26	**0.00**	0.08	*		*		**0.00**	0.40	**0.00**	2.90	2.90
12BRAZIL58	**0.00**	0.78	**0.00**	0.10	*		*		**0.00**	1.40	**0.00**	3.00	3.00
14ST70	**0.00**	0.35	**0.00**	0.07	**0.00**	1.70	**0.00**	0.80	**0.00**	1.20	**0.00**	7.30	7.30
16EIL76	**0.00**	0.37	**0.00**	0.11	**0.00**	2.20	**0.00**	1.10	**0.00**	1.40	**0.00**	9.40	9.40
16PR76	**0.00**	0.45	**0.00**	0.16	**0.00**	2.50	**0.00**	1.90	**0.00**	0.60	**0.00**	12.90	12.90
20RAT99	**0.00**	0.50	**0.00**	0.24	**0.00**	5.00	**0.00**	7.30	**0.00**	3.10	**0.00**	51.40	51.50
20KROA100	**0.00**	0.63	**0.00**	0.25	**0.00**	6.80	**0.00**	3.80	**0.00**	2.40	**0.00**	18.30	18.40
20KROB100	**0.00**	0.60	**0.00**	0.22	**0.00**	6.40	**0.00**	2.40	**0.00**	3.10	**0.00**	22.10	22.20
20KROC100	**0.00**	0.62	**0.00**	0.23	**0.00**	6.50	**0.00**	6.30	**0.00**	2.20	**0.00**	14.30	14.40
20KROD100	**0.00**	0.67	**0.00**	0.43	**0.00**	8.60	**0.00**	5.60	**0.00**	2.50	**0.00**	14.20	14.30
20KROE100	**0.00**	0.58	**0.00**	0.15	**0.00**	6.70	**0.00**	2.80	**0.00**	0.90	**0.00**	12.90	13.00
20RD100	**0.00**	0.51	**0.00**	0.29	0.08	7.30	0.08	8.30	0.08	2.60	**0.00**	16.50	16.60
21EIL101	**0.00**	0.48	**0.00**	0.18	0.40	5.20	0.40	3.00	**0.00**	1.70	**0.00**	25.50	25.60
21LIN105	**0.00**	0.60	**0.00**	0.33	**0.00**	14.40	**0.00**	3.70	**0.00**	2.00	**0.00**	16.20	16.40
22PR107	0.00	0.53	0.00	0.20	**0.00**	8.70	**0.00**	5.20	**0.00**	2.10	**0.00**	7.30	7.40
24GR120	**0.00**	0.66	**0.00**	0.32	*		*		1.99	4.90	**0.00**	41.80	41.90
25PR124	**0.00**	0.68	**0.00**	0.26	0.43	12.20	**0.00**	12.00	**0.00**	3.70	**0.00**	25.70	25.90
26BIER127	**0.00**	0.78	**0.00**	0.28	5.55	36.10	9.68	7.80	**0.00**	11.20	**0.00**	23.30	23.60
28PR136	**0.00**	0.79	0.16	0.36	1.28	12.50	5.54	9.60	0.82	7.20	**0.00**	42.80	43.00
29PR144	**0.00**	1.00	**0.00**	0.44	**0.00**	16.30	**0.00**	11.80	**0.00**	2.30	**0.00**	8.00	8.20
30KROA150	**0.00**	0.98	**0.00**	0.32	**0.00**	17.80	**0.00**	22.90	**0.00**	7.60	**0.00**	100.00	100.30
30KROB150	**0.00**	0.98	**0.00**	0.71	**0.00**	14.20	**0.00**	20.10	**0.00**	9.90	**0.00**	60.30	60.60
31PR152	**0.00**	0.97	**0.00**	0.38	0.47	17.60	1.80	10.30	**0.00**	9.60	**0.00**	51.40	94.80
32U159	**0.00**	0.98	**0.00**	0.55	2.60	18.50	2.79	26.50	**0.00**	10.90	**0.00**	139.60	146.40
39RAT195	**0.00**	1.37	**0.00**	1.33	**0.00**	37.20	1.29	86.00	1.87	8.20	**0.00**	245.50	245.90
40D198	**0.00**	1.63	0.07	1.47	0.60	60.40	0.60	118.80	0.48	12.00	**0.00**	762.50	763.10
40KROA200	**0.00**	1.66	**0.00**	0.95	**0.00**	29.70	5.25	53.00	**0.00**	15.30	**0.00**	183.30	187.40
40KROB200	0.05	1.63	0.01	1.29	**0.00**	35.80	**0.00**	135.20	0.05	19.10	**0.00**	268.00	268.50
45TS225	0.14	1.71	0.28	1.09	0.61	89.00	**0.00**	117.80	0.09	19.40	0.09	1298.40	37875.90
46PR226	**0.00**	1.54	**0.00**	1.09	**0.00**	25.50	2.17	67.60	**0.00**	14.60	**0.00**	106.20	106.90
53GIL262	**0.45**	3.64	0.55	3.05	5.03	115.40	1.88	122.70	3.75	15.80	0.89	1443.50	6624.10
53PR264	**0.00**	2.36	0.09	2.72	0.36	64.40	5.73	147.20	0.33	24.30	**0.00**	336.00	337.00
60PR299	0.05	4.59	0.16	4.08	2.23	90.30	2.01	281.80	**0.00**	33.20	**0.00**	811.40	812.80
64LIN318	**0.00**	8.08	0.54	5.39	4.59	206.80	4.92	317.00	0.36	52.50	0.36	847.80	1671.90
80RD400	**0.58**	14.58	0.72	10.27	1.23	403.50	3.98	1137.10	3.16	59.80	2.97	5031.50	7021.40
84FL417	0.04	8.15	0.06	6.18	0.48	427.10	1.07	1341.00	0.13	77.20	**0.00**	16714.40	16719.40
88PR439	**0.00**	19.06	0.83	15.09	3.52	611.00	4.02	1238.90	1.42	146.60	**0.00**	5418.90	5422.80
89PCB442	**0.01**	23.43	1.23	11.74	5.91	567.70	0.22	838.40	4.22	78.80	0.29	5353.90	58770.50
Averages	**0.03**	2.69	0.11	1.77	0.98	83.09	1.48	171.56	0.46	16.44	0.11	964.79	3356.47
#Trials	5		5		1		1		1		1		1
Platform	Dell Dimension 8400				Sun Sparc Station LX				HP 9000 / 720				

selecting $m = \lceil n/5 \rceil$ centers of clusters such that each center maximizes its distance from the closest already-selected center. Then, all n nodes are added to the cluster whose center is closest.

Computational tests were run on the data. Since Fischetti et al.'s branch and cut (B&C) algorithm provided exact values for TSPLib datasets

* The NN and GI^3 heuristics were not tested in [11] on these datasets.

with size $48 = n = 442$ in [3], direct comparisons with the optimal values were possible on these datasets.

Table 3 follows the format in [14] and provides for each dataset a comparison of percentage above optimal and runtime for the heuristics considered, with bolded entries denoting the best average heuristic solution quality on a dataset. The entries that are not bolded even though they have the value 0.00 indicate that modeling runs were not perfectly optimal, but that the average percentage above optimal rounded down to 0.00. The "Dataset Name" category identifies the name of the dataset considered, with the number of clusters preceding the name and the number of nodes following. For each grouping of columns, "Pct" denotes the average percentage above optimal of the run or runs and "Time" denotes the average runtime of the run or runs, in seconds. The "# Trials" row details the number of trials run per dataset for each algorithm, and the "Platform" row contains the computing platform on which testing was performed. The "GI^3" column refers to Renaud and Boctor's GI^3 heuristic found in [11], the "NN" column refers to Noon's generalized nearest neighbor heuristic followed by GI^3 improvement found in [11], and "FST-Lagr" and "FST-Root" respectively refer to the Lagrangian and root-node heuristics found in [3].

No percentage above optimal was provided for the B&C column, as that algorithm always produces optimal solutions.

The mrOX GA produced, on average, better solution qualities than the other heuristics. Over the datasets considered in Table 3, the mrOX GA averaged only a 0.03% error, less than a third that of the S+D GA and FST-Root heuristic, the two algorithms with the nearest solution qualities. It should be noted that FST-Root had slow runtimes, running within 5% of the exact algorithm's runtime on 35 of the 41 datasets.

The solution qualities produced by the mrOX GA were also close to the published optimal solutions to certain difficult problems being investigated, like 89PCB442, an 89-cluster, 442-node GTSP dataset found in the TSPLib [10]. The algorithm found an optimal solution in four of the five trials run, averaging a 0.01% error over the five trials.

While previous papers presenting heuristics have, in general, limited their scope to problems for which optimal solutions have been published so that percentages above optimal can be calculated, this paper seeks to investigate larger datasets for which the exact algorithm's solution has not been determined due to prohibitively high runtimes. These datasets are clearly the ones for which heuristics are most applicable, and thus should be of the most interest to those who design approximate algorithms.

Thus, five trials were completed comparing the S+D GA and the mrOX GA based on runtime and solution quality on TSPLib datasets of size $493 = n = 1084$, with full results presented in the appendix.

Since no optimal solutions have been published for the larger problems, the success of the mrOX GA was gauged by its performance in relation to the S+D GA on the same datasets. Nearly all mrOX GA solutions to datasets had equal or superior solution qualities compared to those of the S+D GA.

Over all datasets, the mrOX GA provided 0.31% better solutions than the S+D GA, though over the larger datasets (containing more than 442 nodes), the average advantage of the mrOX GA was 1.09%. These are significant improvements, as neither the S+D GA nor the mrOX GA averaged more than 1.09% above optimal for any dataset with 442 or fewer nodes, and the average percentage above optimal for the S+D GA was just 0.11% for the smaller problems. Over the same larger datasets, the mrOX GA produced a better average solution quality than the S+D GA on 12 of the 13 datasets.

The S+D GA, meanwhile, demonstrated on average a 42.79% faster runtime than the mrOX GA. On the larger datasets tested (containing more than 442 nodes), the S+D GA had a 28.79% advantage in runtime, significantly less than the advantage over all datasets, suggesting that the runtimes will continue to remain comparable for larger datasets.

Runtime comparisons with other heuristics were difficult because different computers with various computing powers were used to test the algorithms.

Experimentation was then completed to consider the feasibility of decreasing total runtime of the mrOX GA while maintaining similar solution qualities. Decreasing the number of static generations before termination in the mrOX GA from 150 to 50 provided this effect. Experimentation (with results available in the "50-Gen Value" and "50-Gen Time (ms)" columns of Table 4 in the appendix) demonstrated an overall decrease of 16.51% in runtime, with a decrease of 0.21% in solution quality. The effects were magnified for datasets of size $493 = n = 1084$, with an overall average decrease of 47.52% in runtime and an average decrease of 0.56% in solution quality. Thus, while solution quality, not runtime, was the focus of this paper, the mrOX GA can produce results of reasonable quality very quickly if slightly modified.

Data were collected to quantify the effects of this paper's novel improvements. The population structure involving seven isolated populations, which was used in the mrOX GA, produced 0.04% better solution qualities than the 1-population (standard GA), which was also tested. Considering the small deviation from optimal for the mrOX GA (the average error for mrOX GA solutions on datasets of size $48 = n = 442$ was 0.03%), this represents a significant improvement in solution quality. However, the 20-population model tested was not significantly different from the 7-population scheme, averaging only 0.006% better solution qualities. Thus, limited benefits can clearly be gained through using isolated populations.

Naturally, maintaining more isolated populations caused a longer runtime for the heuristic. For each dataset tested, the 1-population model averaged 43.05 seconds of runtime, the 7-population model averaged 44.44 seconds of runtime, and the 20-population model averaged 49.04 seconds of runtime. As dataset size increases, the percentage of total runtime used in early improvement significantly decreases, from 48.02% for the small 22PR107 to 5.09% for the large 212U1060.

Experimentation was also carried out to determine the advantages of the mrOX crossover over the OX crossover. The mrOX crossover demonstrated a significant advantage in solution quality over the OX crossover, averaging a 0.18% increase in solution quality. The runtimes of the algorithms using the mrOX and OX crossovers were not significantly different, with the mrOX GA running on average 2.59% quicker.

4. CONCLUSIONS

Based on the data collected, the mrOX GA detailed in this paper outperformed all of the other heuristic solutions considered in terms of solution quality, while maintaining comparable runtimes, especially on larger datasets. A trend was established demonstrating an overall improvement in mrOX GA solution qualities in comparison to other heuristics like Snyder and Daskin's GA described in [14]. Additionally, the mrOX GA consistently provided optimal solutions to historically difficult datasets like 89PCB442. It could also be easily modified to provide faster solutions of good (but slightly diminished) quality.

The heuristic thus performed well in comparison to other published algorithms for run-time characteristics, and is further useful because GAs are quite simple to implement in comparison to other heuristics like the FST-Root method. Additionally, changing evaluation functions or performing basic structural transformations into other related problems like the Median Tour Problem described in [1] or Traveling Circus Problem considered in [12] are simple tasks with a GA. However, the effectiveness of these transformations would have to be investigated experimentally.

This paper's research can be applied to other GA solutions of transportation problems through the mrOX crossover. This new crossover significantly improved solution qualities while maintaining similar runtimes in comparison to the OX crossover, characteristics that make it useful in a variety of GAs. Additionally, the initial population isolation mechanism, which was proven to provide better results than a standard GA, can be

Table 4. Experimental Data Collected

Dataset Name	Value	Time (ms)	Merge Time (ms)	Swaps	2-opts	Crossovers	50-Gen Value	50-Gen Time (ms)	S+D GA Value	S+D GA Time (ms)
10ATT48	5394.0	356.0	118.6	3171.2	1592.6	31565.4	5394.0	209.6	5394.0	178.2
10GR48	1834.0	321.8	90.6	2238.8	1601.2	33357.8	1834.0	190.6	1834.0	75.2
10HK48	6386.0	312.8	90.8	3837.6	1102.2	30119.2	6386.0	175.2	6386.0	81.2
11EIL51	174.0	259.2	75.0	1553.2	695.0	23491.8	174.0	134.6	174.0	78.2
11BERLIN52	4040.0	315.4	87.2	2694.8	1490.8	29458.0	4040.0	196.8	4040.0	106.2
12BRAZIL58	15332.0	775.2	228.0	1798.0	1715.4	28722.6	15332.0	190.6	15332.0	97.0
14ST70	316.0	353.0	137.4	1671.4	820.6	26347.8	316.0	225.0	316.0	65.6
16EIL76	209.0	369.0	134.4	1249.2	890.0	26419.8	209.0	228.4	209.0	106.4
16PR76	64925.0	447.0	172.0	3587.2	1293.4	31823.2	64925.0	290.8	64925.0	156.2
20RAT99	497.0	500.0	169.0	2389.8	1437.2	32818.8	497.0	356.2	497.0	243.8
20KROA100	9711.0	628.2	222.0	4876.2	2007.8	39313.8	9711.0	731.2	9711.0	249.8
20KROB100	10328.0	603.2	224.8	4627.0	1682.2	37787.2	10328.0	462.4	10328.0	215.6
20KROC100	9554.0	621.8	206.4	5018.4	2335.4	39378.4	9554.0	443.4	9554.0	225.0
20KROD100	9450.0	668.8	250.0	5078.0	2464.0	39192.8	9450.0	853.0	9450.0	434.4
20KROE100	9523.0	575.0	240.6	3663.8	1721.8	37705.8	9523.0	672.0	9523.0	147.0
20RD100	3650.0	506.2	231.4	2891.6	1292.2	32891.0	3650.0	1003.2	3650.0	290.8
21EIL101	249.0	478.2	218.8	2261.0	1378.2	30152.2	249.0	1434.4	249.0	184.6
21LIN105	8213.0	603.2	256.4	3754.4	1910.0	37844.6	8213.0	1887.2	8213.0	334.4
22PR107	27898.6	534.4	256.6	1340.8	888.8	34388.4	27898.0	1537.4	27898.6	197.0
24GR120	2769.0	659.6	284.4	3176.6	1868.4	37310.2	2769.0	1606.0	2769.0	321.8
25PR124	36605.0	678.0	322.0	3734.8	1626.6	38968.6	36605.0	1118.8	36605.0	259.0
26BIER127	72418.0	784.4	334.4	5499.6	2792.4	40084.2	72418.0	906.4	72418.0	275.2
26CH130	2828.0	790.6	328.2	4126.8	2108.8	43434.2	2828.0	750.4	2828.0	418.4
28PR136	42570.0	793.8	356.2	4200.2	2066.0	38556.8	42570.0	568.8	42639.8	362.8
29PR144	45886.0	1003.2	434.6	5946.6	2776.4	53049.8	45886.0	709.2	45887.4	437.6
30CH150	2750.0	884.4	378.0	4454.6	2743.6	41030.6	2750.0	630.8	2750.0	403.2
30KROA150	11018.0	981.2	421.8	4315.0	2471.0	46399.0	11018.0	621.8	11018.0	319.0
30KROB150	12196.0	978.4	368.8	5270.4	2252.0	45276.6	12196.0	675.2	12196.0	712.4
31PR152	51576.0	965.4	349.8	5753.6	3424.4	39005.6	51577.6	587.6	51576.0	381.2
32U159	22664.0	984.4	381.2	4529.0	2789.8	42891.8	22664.0	675.0	22664.0	553.2
35SI175	5564.0	974.8	353.2	3826.4	3886.2	36402.2	5564.2	806.4	5590.4	387.2
39RAT195	854.0	1374.8	543.8	4307.8	2485.6	50919.2	854.0	868.8	854.0	1325.0
40D198	10557.0	1628.2	572.0	7795.4	3864.4	51261.8	10563.8	996.6	10564.0	1468.6
40KROA200	13406.0	1659.4	590.6	6197.6	3389.6	59078.2	13406.0	1037.2	13406.0	950.2
40KROB200	13117.6	1631.4	618.8	5786.0	2949.6	62330.8	13115.4	1081.4	13112.2	1294.2
45TS225	68435.2	1706.2	593.6	5156.8	3472.8	52474.6	68613.6	1078.0	68530.8	1087.4
46PR226	64007.0	1540.6	712.4	2783.4	2501.2	60787.4	64007.0	968.6	64007.0	1094.0
53GIL262	1017.6	3637.4	912.4	8949.2	5856.4	73077.2	1022.2	1587.6	1018.6	3046.8
53PR264	29549.0	2359.4	1012.6	4445.4	2638.2	71733.4	29549.0	1475.0	29574.8	2718.6
56A280	1080.8	2921.8	1018.8	5591.8	3314.8	68932.6	1088.8	1806.2	1080.6	3321.8
60PR299	22627.0	4593.8	1415.8	8194.2	4062.8	92713.4	22647.2	3540.8	22650.2	4084.4
64LIN318	20765.0	8084.4	1475.2	16282.6	7666.8	91508.2	21036.6	3565.6	20877.8	5387.6
80RD400	6397.8	14578.2	2453.2	19330.2	7989.2	117979.2	6413.2	8041.0	6407.0	10265.6
84FL417	9654.6	8152.8	2312.2	6724.2	5790.8	110035.2	9668.2	4553.4	9657.0	6175.2
88PR439	60099.0	19059.6	3581.6	19792.0	8235.8	143845.4	60348.2	10996.6	60595.4	15087.6
89PCB442	21658.2	23434.4	3309.4	26512.2	12235.8	137437.0	21904.0	10927.8	21923.0	11743.8
99D493	20117.2	35718.8	3675.0	33168.8	13203.8	134546.2	20146.2	21972.0	20260.4	14887.8
107ATT532	13510.8	31703.0	4440.4	24098.0	10421.6	145720.0	13520.0	20043.6	13529.8	31875.2
107SI535	13513.2	26346.8	3518.8	16626.2	19904.8	118799.6	13533.2	14543.8	13557.6	11250.2
113PA561	1053.6	21084.2	3837.6	10258.4	11026.2	127844.0	1051.2	13759.4	1065.6	26818.6
115RAT575	2414.8	48481.0	5706.0	29366.8	12684.8	177752.8	2436.4	23506.2	2442.4	46834.4
131P654	27508.2	32672.0	5909.4	11381.4	10344.2	173235.6	27439.0	17903.0	27448.4	46996.8
132D657	22599.0	132243.6	8681.2	66083.8	22186.6	218083.4	22624.0	59046.8	22857.6	58449.8
145U724	17370.6	161815.2	9921.6	62581.8	22077.2	223626.6	17681.8	58994.0	17806.2	59625.2
157RAT783	3300.2	152147.0	12421.8	48479.2	19752.6	231742.4	3330.8	68056.2	3341.0	89362.4
201PR1002	114582.2	464356.4	26940.6	89278.4	27910.8	339429.2	116058.4	295209.2	117421.2	332406.2
207SI1032	22388.8	242366.0	19397.2	32418.6	52047.0	253389.2	22415.2	126962.4	22515.2	135431.0
212U1060	108390.4	594637.4	30281.4	106204.4	30766.4	352773.6	109519.8	239453.2	110158.0	216999.8
217VM1084	131884.6	562040.6	32193.6	78499.2	32331.6	331020.4	133563.4	290765.6	133743.4	390115.6

applied to a wide variety of GAs. This is a far-reaching application of this paper's findings, considering the many GAs used as heuristic solutions in computing today.

Further research could be conducted on the effects of the novel improvements on the schemata theorem, the basic theoretical support for GAs. Additional research could consider the use of an entirely different crossover (such as the edge recombination crossover) in conjunction with a rotational inversion mechanism, or the effectiveness of a slightly modified mrOX GA on other transportation problems.

5. APPENDIX

In Table 4, the "Value" column contains the mrOX GA fitness value, the "Time (ms)" column contains the total mrOX GA runtime including time both before and after the population merge, the "Merge Time (ms)" column contains the mrOX GA's runtime until it merges isolated populations, the "Swaps" column contains the mrOX GA number of swaps, the "2-opts" column contains the number of mrOX GA 2-opts, the "Crossovers" column contains the number of mrOX GA crossovers, the "50-Gen value" column contains the mrOX GA fitness value for the 50-generation termination run, the "50-Gen Time (ms)" column contains the total mrOX GA runtime for the 50-generation termination run, the "S+D GA Value" column contains the S+D GA fitness value, and the "S+D GA Time (ms)" column contains the S+D GA's runtime. S+D values and runtimes are from this paper's coding of the heuristic. Fractional values are the effects of averaging results from 5 trial runs.

6. REFERENCES

1. J.R. Current, D.A. Schilling, 1994, The median tour and maximal covering tour problems: Formulations and heuristics. *European Journal of Operational Research,* **73**: 114–126.
2. L. Davis, 1985, Applying Adaptive Algorithms to Epistatic Domains. *Proceeding of the International Joint Conference on Artificial Intelligence*, 162-164.
3. M. Fischetti, J.J. Salazar-Gonzalez, P. Toth, 1997, A branch-and-cut algorithm for the symmetric generalized traveling salesman problem. *Operations Research,* **45**(3): 378–394.
4. D.E. Goldberg and R. Lingle, 1985, Alleles, loci and the traveling salesman problem. In: *J.J. Grefenstette (ed.), Proceedings of the First International Conference on Genetic Algorithms and Their Applications*, 154-159, Lawrench Erlbaum Associates, Hillsdale, N.J.

5. A.L. Henry-Labordere, 1969, The record balancing problem: A dynamic programming solution of a generalized traveling salesman problem. *RAIRO,* **B2**: 43-49.
6. G. Laporte, A. Asef-Vaziri, C. Sriskandarajah, 1996, Some Applications of the Generalized Traveling Salesman Problem. *Journal of the Operational Research Society,* **47**: 1461-1467.
7. Z. Michalewicz, 1999, Genetic Algorithms + Data Structures = Evolution Programs. Springer-Verlag, Charlotte, NC.
8. H. Mühlenbein, M.G. Schleuter, and O. Krämer, 1988, Evolution algorithms in combinatorial optimization. *Parallel Computing,* **7**: 65-85.
9. C.E. Noon, 1988, The generalized traveling salesman problem. Ph. D. Dissertation, University of Michigan.
10. G. Reinelt, 1996, TSPLIB—A traveling salesman problem library. *ORSA Journal on Computing,* **4**: 134–143.
11. J. Renaud, F.F. Boctor, 1998, An efficient composite heuristic for the symmetric generalized traveling salesman problem. *European Journal of Operational Research,* **108**(3): 571–584.
12. C.S. Revelle, G. Laporte, 1996, The plant location problem: New models and research prospects. *Operations Research,* **44**(6): 864–874.
13. J.P. Saksena, 1970, Mathematical model of scheduling clients through welfare agencies. *CORS Journal,* **8**: 185-200.
14. L. Snyder and M. Daskin, 2006, A random-key genetic algorithm for the generalized traveling salesman problem. *European Journal of Operational Research,* **17**(1): 38-53.
15. H.-K. Tsai, J.-M. Yang, Y.-F. Tsai, C.-Y. Kao, 2004, Some issues of designing genetic algorithms for traveling salesman problems. *Soft Computing,* **8**: 689-697.

SENSITIVITY ANALYSIS IN SIMULATION OF STOCHASTIC ACTIVITY NETWORKS: A COMPUTATIONAL STUDY

Chris Groër[1] and Ken Ryals[2]
[1]*University of Maryland; and* [2]*Johns Hopkins University Applied Physics Laboratory*

Abstract: Two important performance measures related to Stochastic Activity Networks (SANs) are the length of the longest path and the probability that this longest path length exceeds a given threshold. We examine the sensitivity of these performance measures to changes in the underlying parameters of the arc length distributions by calculating four different derivative estimators via Monte Carlo simulation. We explore the statistical properties of these estimators and suggest a method of combining these estimators as a tool for variance reduction

Key words: Stochastic Activity Network, sensitivity, variance reduction, derivative estimators

1. INTRODUCTION

A Stochastic Activity Network (SAN) is a directed acyclic graph where the lengths of the arcs are random variables. SANs have a wide range of modeling applications, from project management to the analysis of complex communication systems. Typically, one is interested in various properties of the paths that connect the source node to the sink node. A particularly important performance measure is the longest path through the network. In the context of the Project Evaluation Review Technique (PERT), for example, the length of the longest path typically represents the total project completion time.

In this paper, we assume that the lengths of the individual arcs in the SAN are governed by probability distributions and we attempt to measure the sensitivity of the length of the longest path to changes in the parameters of these underlying distributions. From the perspective of a project manager, such sensitivity estimates provide information on how best to distribute constrained resources to certain tasks. For example, after determining that the total project completion time is particularly sensitive to the probability distribution parameter of a particular job, additional resources can be expended on this task, thereby changing the relevant parameter and reducing the total project completion time. This type of "time-cost tradeoff" is discussed in more detail in [1].

Because of the complexity encountered even in very small networks, estimating this sensitivity requires a simulation-based approach. Several different techniques have been developed for computing these types of sensitivity estimates [6]. A summary of such estimators is found in Elmaghraby's paper [3], and we take the later work of Fu [4] as the starting point of our computational study. We empirically investigate five different techniques for estimating the derivative of a performance measure with respect to parameters defining individual components:

- Finite difference (FD) techniques,
- Likelihood ratio (LR) method – also called the Score Function (SF) method,
- Weak derivative (WD) estimators,
- Infinitesimal perturbation analysis (IPA), and
- Smoothed perturbation analysis (SPA) extension to IPA.

These methods will be applied to two performance measures related to the longest path through the SAN. The first performance measure is the length of the longest path through the network, (i.e. the "critical" path), and the second is the probability that the longest path length will exceed a threshold value. We compute the simulation-based derivative estimators for two different networks and empirically compare the performance of the different methods.

We explore the impact of using common random numbers (CRN) for the FD and WD estimators. For the two metrics examined, we explore relationships between the different estimation techniques and present a potential improvement to the overall estimate.

2. PRELIMINARIES

Given a particular SAN with n nodes and k edges, we label these nodes as $\{1,2,\ldots,n\}$, taking node 1 as the source and node n as the sink, and label the k arcs of our SAN as $\{1,2,\ldots,k\}$. We begin by presenting a small five-node SAN with six arcs. This SAN is shown in Figure 1 and is taken directly from [4]. There are three paths from source to sink:

- Node 1→ Node 2→ Node 4→ Node 5: (Arcs 1, 4, and 6),
- Node 1→ Node 3→ Node 4→ Node 5: (Arcs 2, 5, and 6),
- Node 1→ Node 2→ Node 3→ Node 4→ Node 5: (Arcs 1, 3, 5, and 6).

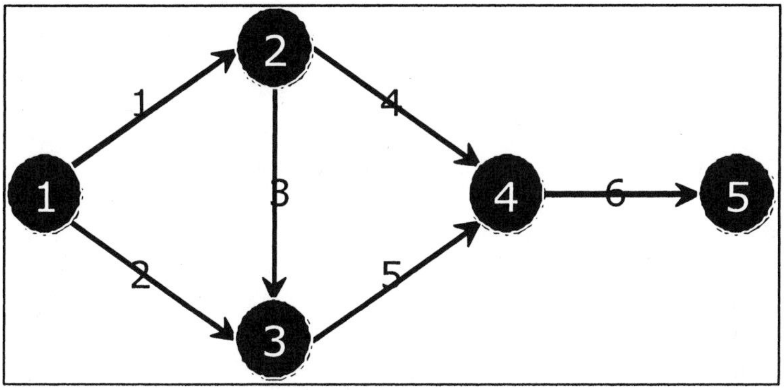

Figure 1. Sample small SAN

Denoting the time to complete the activity represented by arc i as X_i, the time to complete a particular path P is Y_P, given by:

$$Y_P = \sum_{i \in P} X_i \ .$$

In this simple case, the longest path P* is simply the longest of the three different paths. Letting Y denote the length of this longest path, the likelihood that the project will be late in completion is represented by P(Y > y), where y is some threshold value. Assume that the length of Arc 1 has probability density function f_1 with parameter $?_1$ and that the individual arc lengths are the random variables $X_1, X_2, \ldots, X_6$. Finally, let Y(X) be the random variable that is the length of the longest path P* under these conditions.

With this notation in mind, we now present a description of the various derivative estimators that we implemented. The simplest derivative estimator is the finite difference estimate, obtained from the definition of a derivative:

$$\frac{dY(X(\theta_1))}{d\theta_1} = \lim_{h \to 0} \frac{Y(X((\theta_1 + h)) - Y(X((\theta_1))}{h} \ .$$

Thus, one can estimate the derivative using an extra simulation (*for each parameter*) with the parameter slightly modified. For a large network, where many sensitivities are of interest, this could involve a large number of additional simulations.

This leads us to the other sensitivity estimators referenced earlier. Since the focus of this paper is computational in nature, we will not derive these other three estimators here (see [4] and [5] for details on the derivations). However, it is worthwhile to look at each estimator in a bit of detail for a simplified case. The IPA estimator for the sensitivity of the longest path length to changes in $?_1$ is given by:

$$\frac{\partial Y}{\partial \theta_1} = \frac{\partial X_1}{\partial \theta_1} \bullet 1\{Arc1 \in P^*\}$$

where **1{*}** denotes the indicator function. From this, we see that it is very simple to compute in simulation as it involves essentially one task—calculating the longest path and determining whether Arc 1 is on this path. For most commonly encountered distributions, the quantity $dX_1/d?_1$ is also very simple to compute, making the IPA estimator especially easy to implement.

The LR estimator takes the form:

$$\frac{\partial Y}{\partial \theta_1} = Y(X) \bullet \frac{\partial \ln(f_1(X_1, \theta))}{\partial \theta_1} .$$

One must again compute the longest path through the network as in the case of the IPA estimator, and the computational expense is only marginally greater than what is required for the IPA estimator, due to a bit more complexity in the second term.

The final estimator we consider here, the WD estimator, is a bit more complex. Let $(c_1(?_1), f_1^{(1)}, f_1^{(2)})$ denote a so-called "weak derivative" for the probability density function f_1 of Arc 1. Then, let $Y(X_1^{(1)})$ be the longest path through the network after we replace the random variable X_1 with a new random variable from the distribution $f_1^{(1)}$ and let $Y(X_1^{(2)})$ be the longest path through the network after we replace X_1 with a random variable from the distribution $f_1^{(2)}$. The pdf's $f_1^{(1)}$ and $f_1^{(2)}$ are chosen so that their difference (normalized by $c_1(?_1)$) is a weak derivative of the original pdf, f_1, for Arc 1 when integrated in an expectation operation. The WD estimator is then calculated as:

$$\frac{\partial Y(X_1)}{\partial \theta_1} = c_1(\theta_1) \bullet \left(Y(X_1^{(2)}) - Y(X_1^{(1)}) \right) \quad X_1^{(j)} \sim f_1^{(j)} \ ,$$

where $\left(c_1(\theta_1), f_1^{(2)}, f_1^{(1)}\right)$ *is a weak derivative for* f_1 *with respect to* θ_1 .

We see that the calculation of this estimator is in general a bit more involved than the first two, since we now must potentially compute the longest paths through two slightly different networks.

3. COMPARING THE PERFORMANCE OF THE DIFFERENT ESTIMATORS

The Finite Difference, IPA, LR, and Weak Derivative Estimators each have their own advantages and disadvantages. Depending on the input distributions of the arc lengths, one or more of these estimators may not be available in all cases. Thus, it is useful to study some of the properties of these estimators for particular input distributions. We did so through numerous Monte Carlo simulations that we performed using the random number generators provided by both *Mathematica* and MATLAB.

We begin by looking at a very simple case addressed in [4] where all arc lengths in the small SAN of Figure 1 are ~exp(1) random variables, and we are estimating the sensitivity of the longest path length to changes in the parameter of the distribution for Arc 1. We will conduct many runs of the simulation and then compare histograms of the different estimators to get a sense of their performance. In this case, we also experimented with the most intuitive of all sensitivity estimators, the finite difference estimator, both with and without common random numbers (CRN).

The first plot (Figure 2) shows the observed densities of the LR, WD, and IPA estimators. We also implemented the finite difference estimator using CRN and found it to be essentially indistinguishable from the IPA estimator. The "spikes" at the value of zero are easily explained by a simple analysis of the estimator function. The IPA estimator is zero when the Arc 1 is not on the critical path and the WD estimator is zero when the length of the critical path remains unchanged after we replace X_1 with random variables drawn from the distributions determined by the weak derivative (Erlang in this particular case).

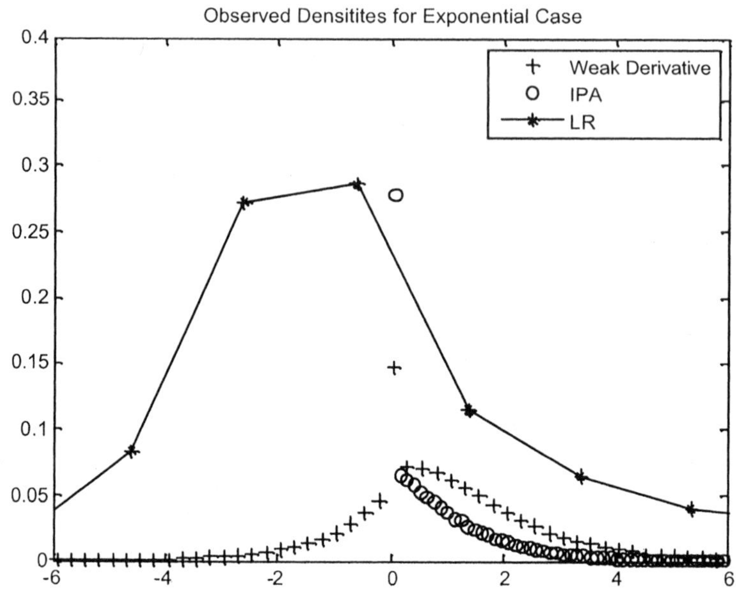

Figure 2. Densities of the estimators for the longest path in the small SAN with arc lengths distributed as ~exp(1)

We see that the IPA estimator has the lowest variance and never takes on negative values. The WD estimator has a very different distribution, as it can take on negative values and has a much higher variance. The distribution of the LR estimator is clearly very different from the others, and we see in Table 1 that its variance tends to be dramatically larger than the variance of the other estimators.

Table 1. Comparison of the estimators for longest time on the small SAN with arc lengths distributed ~exp(1)

Estimator	**Mean**	**Standard Deviation**	**95% Confidence Interval**
LR	0.901	6.87	(0.870,0.931)
IPA	0.895	1.05	(0.890,0.899)
WD without CRN	0.899	1.66	(0.892,0.907)
FD with CRN	0.893	1.04	(0.888,0.897)
FD without CRN	1.387	132	(0.808,1.965)

Although the simple-minded finite difference method performs poorly when we do not use CRN, this estimator appears to be as good as any other when CRN is implemented, as the standard deviation of this estimator is lower than any other estimator.

A second input distribution that we considered is the case where all arcs are uniformly distributed. In particular, for the small SAN, we let all arcs be distributed as U[0,2] except for Arc 1, which has the distribution U[1-$?_1$,1+$?_1$]. The estimators are then measuring sensitivity of the longest path length to changes in this parameter $?_1$. We found that the finite difference estimator was virtually indistinguishable from the IPA estimator. As the LR estimator is undefined in this case, the density plot in Figure 3 shows only the WD and IPA estimators for the case $?_1$ =1.00.

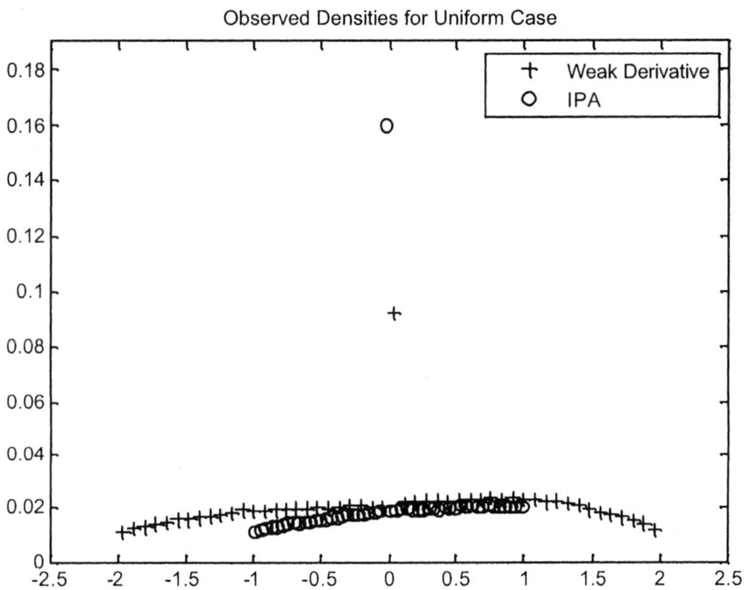

Figure 3. Densities of the estimators for longest path on the small SAN with arc lengths distributed as ~U[0,2]

If we consider FD without CRN, we find that it once again has a very large variance compared with the other estimators as shown in Table 2.

Table 2. Comparison of the estimators for longest path on the small SAN with arc lengths distributed ~U[0,2]

Estimator	Mean	Standard Deviation	95% Confidence Interval
LR		*Not Applicable*	
IPA	0.074	0.52	(.072,.076)
WD without CRN	0.074	1.03	(.069,.079)
FD with CRN	0.076	0.52	(.074,.078)
FD without CRN	0.305	74.2	(-.020,.630)

The results are quite similar to those obtained for the exponential distribution: IPA and FD with CRN are both easy to compute (provided synchronization is not a difficulty) and both have a low variance. The WD estimator is not too far behind, but FD without CRN has a very large variance once again. It is interesting to note that in this case, the sensitivity of the critical path length to $?_1$ is significantly lower than when all arcs are exponentially distributed. This makes intuitive sense since when we vary $?_1$ in this case we are not changing the mean length of Arc 1, just its variance. This is in contrast to the exponential case where varying the parameter $?_1$ also changes the mean.

Similar experiments were carried out for other combinations of distributions on this small SAN. In the first combination, the lengths of Arcs 2-6 are ~gamma(2, ½) and Arc 1 is ~gamma(2, $?_1$). We are measuring the sensitivity of the length of the longest path to changes to the parameter ? when $?_1$= ½. In the second combination, Arcs 2-6 are ~U(0,2) and Arc 1 is again ~gamma(2, $?_1$). In both cases, the performance of the different sensitivity estimators was always very similar to what we observed earlier. These results for additional distributions are summarized in Tables 3 and 4.

Table 3. Comparison of the estimators for longest path on the small SAN with arc lengths distributed ~Gamma[2, ½]

Estimator	Mean	Standard Deviation	95% Confidence Interval
LR	1.809	15.411	(1.741,1.877)
IPA	1.814	1.545	(1.807,1.821)
WD without CRN	1.834	4.270	(1.815,1.853)
FD with CRN	1.815	1.546	(1.808,1.821)
FD without CRN	2.139	94.793	(1.130, 1.962)

Table 4. Comparison of the estimators for longest path on the small SAN with Arc 1 ~Gamma[2, ½] and all other arc lengths ~U[0,2]

Estimator	Mean	Standard Deviation	95% Confidence Interval
LR	1.797	15.21	(1.730,1.864)
IPA	1.838	1.545	(1.831,1.844)
WD without CRN	1.829	4.309	(1.810,1.848)
FD with CRN	1.849	1.539	(1.842,1.856)
FD without CRN	2.368	95.195	(1.538,2.372)

The trend is obvious: IPA and FD with CRN consistently outperform the WD and LR estimators in terms of having the lowest variance. Thus, these two estimators should be the first choices when they are available in these types of simulation problems.

We conclude our comparison of the different sensitivity estimators by applying them to a substantially larger network, taken directly from [3], shown in Figure 4. In this 20 node network, there are 51 paths from source to sink using 38 arcs.

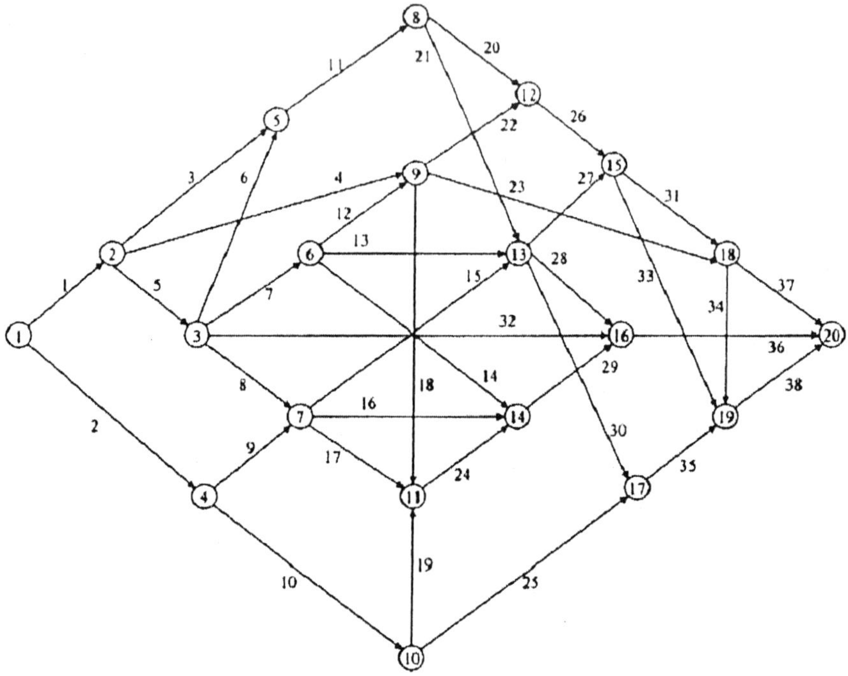

Figure 4. Sample large SAN

The conclusions regarding relative performance of the estimators on the small SAN was the same for all of the arc length distributions; thus, we restricted our simulation to the case where all arcs are ~exp(1) random variables. We measured the sensitivity of the expected longest path length to changes in the parameter for the input distribution of Arc 1. We implemented all four estimators and used CRN for the finite difference estimator. Due to the poor performance of the FD estimator without CRN on the small SAN, it was not examined for the large SAN. Just as in the case of the smaller SAN studied earlier, we found the IPA and CRN finite difference estimators to be indistinguishable. We again ran the simulation 200,000 times. In Figure 5, we display a histogram of the observed densities of the various estimators.

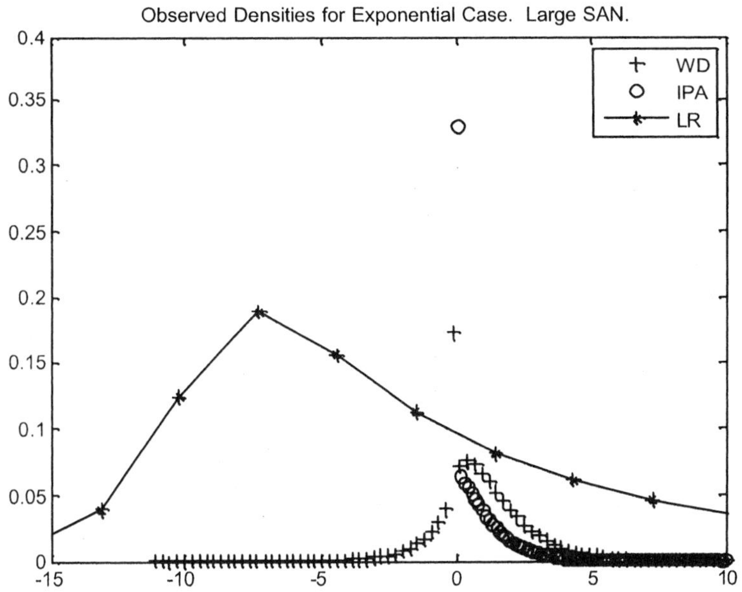

Figure 5. Densities of the estimators for longest path on the large SAN with arc lengths distributed ~exp(1)

The plot of these observed densities is quite similar to what we observed for the smaller SAN. Table 5 shows that the performance of the different estimators is comparable to what we observed earlier for the small SAN. This is not surprising. However, it is important to notice that the variance of the LR estimator has increased by roughly a factor of four. This is due to the presence of the term Y(X) in the LR estimator, which causes the variance to grow as the length of the longest path, grows. Loosely speaking, then, the larger the network, the larger the variance for the LR estimator.

Table 5. Comparison of the estimators for longest path on the large SAN with arc lengths distributed ~exp(1)

Estimator	Mean	Standard Deviation	95% Confidence Interval
LR	0.901	14.29	(0.838,0.964)
IPA	0.929	1.02	(0.925,0.933)
WD without CRN	0.936	1.69	(0.929,0.943)
FD with CRN	0.934	1.03	(0.929,0.939)

A second performance measure that is discussed in [4] is P(Y>y), the tail distribution of the longest path length, which is the probability that the longest path length Y exceeds some threshold. We consider the small SAN with ~exp(1) arc lengths and consider the sensitivity of P(Y>y) to changes in the parameter ? of Arc 1. In this case, three sensitivity estimators can be calculated for this performance measure: LR, WD, and Smoothed Perturbation Analysis (SPA). IPA is not available here due to the discontinuity of the indicator function. The LR and WD estimators are obtained simply by replacing Y with **1**{Y>y} in the formulas given earlier. The derivation of the SPA estimator is significantly more complicated as it requires one to compute the derivative of a *conditional* expectation. Here, we condition on the set of all arc lengths except X_1 and we must compute various sums of the other arc lengths to compute the estimator. The SPA estimator can be computed as follows (see Fu[4] for more details):

$$Let\ \tilde{y} = y - MAX(Arc3 + Arc5, Arc4 - Arc6)\ and,$$

$$M = Max(Arc3 + Arc5 + Arc6, Arc4 + Arc6, Arc2 + Arc5 + Arc6)$$

$$\frac{dY}{d\theta_1} = \frac{\tilde{y}}{\theta_1^2} e^{-\frac{\tilde{y}}{\theta_1}} \bullet 1\{M \leq y\}$$

For our computational example, we chose y=4.0. Note that in this case, the Weak Derivative Estimator can assume only the values -1, 0, and 1. Thus, its histogram is not particularly interesting so we omit it from the plot below. After removing the "spikes" that occur at 0 for each estimator, we have the histogram of Figure 6 showing the observed values of the LR and SPA estimators. We found the SPA distribution quite unusual due to its odd shape. While this distribution has quite a bit of mass at the right hand side of the distribution, the spike at the value of 0 offsets this and provides an estimator with a very low variance in this case. As before, we used 200,000 runs to generate our datasets for the histograms.

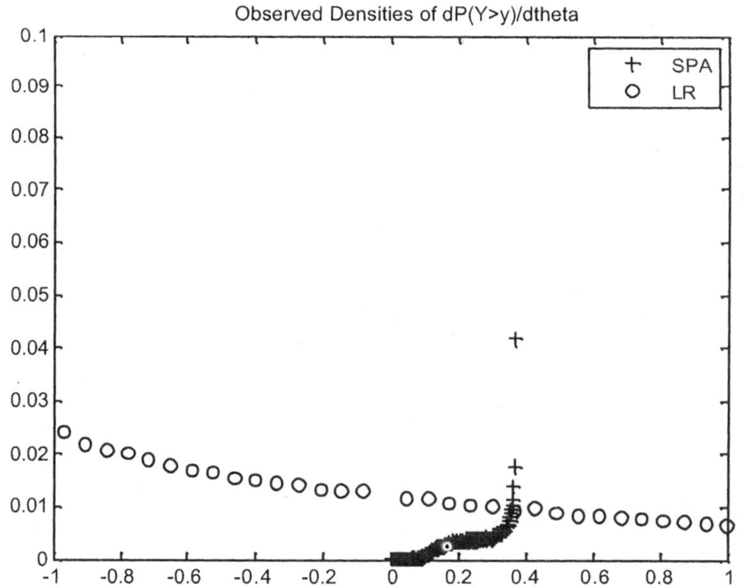

Figure 6. Portion of the estimator densities for *d*P(Y>*y*)/*d*?$_1$ on the small SAN with arc lengths distributed ~exp(1)

As before, we provide Table 6 which compares various statistical properties of these three estimators.

Table 6. Comparison of the estimators for *d*P(Y>*y*)/*d*?$_1$ on the small SAN with arc lengths distributed ~exp(1)

Estimator	Mean	Standard Deviation	95% Confidence Interval
LR	0.1770	0.8851	(0.173,0.181)
SPA	0.1718	0.1446	(0.171,0.172)
WD without CRN	0.1699	0.4841	(0.168,0.172)

Based on this example, although the SPA estimator requires more care in its derivation and implementation, it appears to perform substantially better than the other estimators as its variance is substantially lower than that of the other estimators.

4. RELATIONSHIPS BETWEEN THE DERIVATIVE ESTIMATES

This section investigates relationships between different estimates for the derivatives. The relative behavior of the two metrics (longest path and likelihood of exceeding a threshold) is examined for the applicable estimators.

For each of the cases above, twenty replications of 5000 trial Monte Carlo simulations were generated using the small SAN with exponential distributions for each arc (with mean=1). For reference, the simple statistics for the derivative estimates are shown in Table 7. One thing that is apparent is that the WD estimator does not suffer very much from an inability to use CRN as the FD estimator does. While the FD estimates without CRN have standard deviations that are approximately 250 and 16 times larger than their CRN counterparts, the WD estimators only exhibit standard deviation increases of 1.6 and 1.2 for derivatives of E(Y) and P(Y>y), respectively.

Table 7. Simple statistics for 20 sets of derivative estimates for longest path on the small SAN with arc lengths ~exp(1)

Derivative Estimation Technique	**Metric**	
	$\frac{dY}{d\theta_1}$	$\frac{dP(Y > y)}{d\theta_1}$
PA (IPA or SPA)	Mean=0.8954 Std Dev=0.0143	Mean=0.1941 Std Dev =0.0020
LR	Mean=0.8849 Std Dev =0.0985	Mean=0.1717 Std Dev =0.0119
WD with CRN	Mean=0.8934 Std Dev =0.0246	Mean=0.1694 Std Dev =0.0069
WD without CRN	Mean=0.9002 Std Dev =0.0401	Mean=0.1708 Std Dev =0.0085
FD with CRN	Mean=0.8958 Std Dev =0.0142	Mean=0.1550 Std Dev =0.0587
FD without CRN	Mean=0.5983 Std Dev =3.6845	Mean=0.0210 Std Dev =0.9352

The relationships for individual pairs of estimates for the longest path metric are summarized in Table 8 in terms of the slope and correlation coefficient. The significance of the slope is in the sign; although most of the estimators are positively correlated, all are negatively correlated with the WD estimate for these random variates. The correlation coefficient indicates that some of the relationships are very strong (the non-CRN results produced approximately zero correlation and a zero slope as expected and are not shown), which indicates that they are measuring both the same metric, and the same noise. Thus, there is potential for combining estimators to reduce variance.

Table 8. Summary of relationships between derivative estimates for longest path on the small SAN with arc lengths ~exp(1)

Slope / Correlation	WD with CRN	LR	IPA
FD with CRN	-1.4432 0.6945	6.2649 0.8160	1.0034 0.9996
WD with CRN		-2.9849 0.5555	-0.485 0.7003
LR			0.131 0.8230

The relationships between the derivative estimates for the probability of exceeding a threshold, P(Y>*y*), are summarized in Table 9. This metric does not exhibit the strong correlations between estimates that the longest path metric did. In fact, the best correlation for this metric is lower than the worst for the longest path derivative. Furthermore, many of the slopes are essentially zero, confirming that there is no meaningful relationship between the derivative estimates for a probability metric.

Table 9. Summary of Relationship Between Derivative Estimates for P(Y>*y*) on the small SAN with arc lengths ~exp(1)

Slope / Correlation	WD with CRN	LR	SPA
FD with CRN	-0.0502 0.1824	0.0333 0.0270	-0.0096 0.0775
WD with CRN		-0.6225 0.1304	0.2006 0.4625
LR			-0.0394 0.0531

Several of the derivative estimates for the expected value of the longest path through the network exhibit significant correlation. Since some pairs, such as LR and PA, tend to occur together, this correlation can be used to produce improved estimates. Consider the LR and IPA estimates for the derivative of longest path. If a weighted average were formed where

$$New\_Estimate = \alpha * LR\_Estimate + (1-\alpha) * IPA\_Estimate,$$

then the variance of the new estimate is:

$$Var(New) = \alpha^2 * Var(LR) + (1-\alpha)^2 * Var(IPA) + 2 * \alpha * (1-\alpha) * CoVar(IPA, LR).$$

The value of a that minimizes the variance of the new estimate is

$$\alpha = \frac{1}{2} * \left(\frac{2 * Var(IPA) - CoVar(IPA, LR)}{Var(LR) + Var(IPA) - CoVar(IPA, LR)} \right).$$

This permits the highly variable LR estimates to improve the much better IPA estimates. For example, consider the following situation generated from a single 5000 trial simulation:

IPA	LR
$Mean = 0.9022$	$Mean = 0.9085$
$Variance = 1.0909$	$Variance = 46.7928$
$CoVariance(IPA, LR) = 6.3380$	
	$New Estimate$
$\hat{\alpha} = -0.0500 \Rightarrow$	$Mean = 0.9019$
	$Variance = 0.6540$

The negative estimate for a results from the positive correlation between the IPA and LR estimates in the example above. Thus, in this example, LR estimates with confidence intervals significantly larger than those for the IPA estimates produced a new set of derivative estimates for which the confidence interval is much smaller than the IPA confidence interval. This "free" improvement was evident for both the small and large SANs and warrants always generating the LR derivative estimates in conjunction with the "better" IPA estimates as a way of reducing the size of generated confidence intervals without additional simulations. Interestingly, this value for a appears to be consistent for different replications as shown in Table 10. In this table, five different 5000 samples sets of the small SAN were generated and the estimate for a generated for each. Then, each value of a was applied to all five sets to determine the degree of variance reduction generated. The small variations in a, coupled with the fact that all of the hybrid sets have a variance of approximately 30% of the (smaller) IPA variance from which they were derived, indicates that the hybrid IPA/LR estimator's performance is not particularly sensitive to the parameter a.

Table 10. Robustness of hybrid estimator to the parameter a

	Set #1	Set #2	Set #3	Set #4	Set #5
LR Mean	0.9085	0.9702	0.7259	0.7299	0.9339
LR Variance	46.7928	53.6894	40.1980	42.2630	45.1779
IPA Mean	0.9022	0.9021	0.8738	0.8717	0.9022
IPA Variance	1.0909	1.1401	1.0188	1.0395	1.0578
Covar (IPA,LR)	6.3380	6.9297	5.7368	5.9476	6.1322
Estimate for a	-0.1490	-0.1413	-0.1586	-0.1563	-0.1494
a used	**Variance of Hybrid (LR and IPA) Estimate**				
-0.1490	0.3089	0.3244	0.2732	0.2742	0.2999
-0.1413	0.3110	0.3220	0.2794	0.2796	0.3021
-0.1586	0.3121	0.3342	0.2705	0.2727	0.3028
-0.1563	0.3107	0.3311	0.2706	0.2726	0.3015
-0.1494	0.3089	0.3246	0.2730	0.2741	0.2999

5. RECOMMENDATIONS AND CONCLUSIONS

We have performed extensive simulations of two different stochastic activity networks in order to understand and study sensitivity estimators. The IPA, LR, and WD estimators provide an efficient and usually superior alternative to the much better known finite difference estimator. Our simulations have allowed us to gain a better understanding of the performance of these estimators in terms of their variances and distributions. When available for a given input distribution, IPA and WD seem to provide the lowest variance estimate, with FD using CRN also an attractive alternative. The LR and FD without CRN have much larger variance. The WD estimator performs nearly as well without CRN as it does with CRN, making it a worthy candidate for cases wherein CRN is not possible. Our study of the sensitivity of the tail distribution led us to the SPA estimator that proved to be surprisingly effective in the case that we considered. Though its density is quite unusual, it exhibited a much lower variance than both the WD and LR estimators did. Finally, we studied the relationships among these different estimators when all arcs are exponentially distributed. By exploiting certain correlations among these estimators, we were able to provide a hybrid estimator that combined different estimators in order to produce a sensitivity estimator for the expected longest path that has an overall lower variance.

ACKNOWLEDGEMENTS

The authors would like to thank the reviewers for their suggestions and Dr. Michael C. Fu for his helpful correspondence.

REFERENCES

[1] R.A. Bowman. Stochastic gradient-based time-cost tradeoffs in PERT networks using simulation. *Annals of Operations Research* 53, 533-551, 1994.

[2] R.A. Bowman. Efficient estimation of arc criticalities in stochastic activity networks. *Management Science* 41, 58-67, 1995.

[3] S.E. Elmaghraby. On criticality and sensitivity in activity networks. *European Journal of Operational Research* 127, 220-238, 2000.

[4] M.C. Fu. Sensitivity Analysis for Simulation of Stochastic Activity Networks, *Topics in Modeling, Optimization, and Decision Technologies: Honoring Saul Gass' Contributions to Operations Research* (tentative title), F.B. Alt, M.C. Fu and B.L. Golden, editors, Kluwer Academic Publishers, 2006

[5] M.C. Fu. Stochastic Gradient Estimation, Chapter 19 in *Handbooks in Operations Research and Management Science: Simulation,* S.G. Henderson and B.L. Nelson, eds., Elsevier, 2006.

[6] Krivulin, Nikolai. Unbiased Estimates for Gradients of Stochastic Network Performance Measures. *Acta Applicandae Mathematicae,* 1993, Vol. 33, p. 21-43.

COMBINED DISCRETE-CONTINUOUS SIMULATION MODELING OF AN AUTONOMOUS UNDERWATER VEHICLE

Roy Jarnagin and Senay Solak
Department of Industrial Engineering Technology, Southern Polytechnic State University
1100 South Marietta Parkway, Marietta, GA 30060

Abstract: In this study, we develop a combined discrete-continuous simulation model for the trajectory control of an autonomous underwater vehicle. The differential equations governing the movements of the autonomous underwater vehicle are integrated numerically and the results are used to simulate the trajectory control of the vehicle. The developed model is to be used in the design of the command and control software and the internal communication interface architecture for the vehicle. The model may also be used to assess the effects of random message delivery variation and sensor inaccuracies.

Key words: discrete-continuous simulation, autonomous underwater vehicle, AUV

1. INTRODUCTION

In combined discrete-continuous simulation models, continuous state variables interact in complex or unpredictable ways with discrete time-stepped events. Typically, the introduction of continuous variables into a discrete event simulation serves to evaluate the continuous variable itself often by numerical integration of a governing differential equation. On the other hand, there is a very large body of literature on the simulation of automatic control for underwater vehicles [1-5]. In many cases the specifics of the simulation environment are not described. In cases where the environment is described, it typically involves multiple interacting software modules running under a real-time operating system. Trajectory control of open-frame type vehicles is especially challenging due to the significant hydrodynamic effects that are generally difficult to characterize accurately. Therefore, most studies include development of complex simulators, in which the hardware and software for the modeled system are connected to the simulation models.

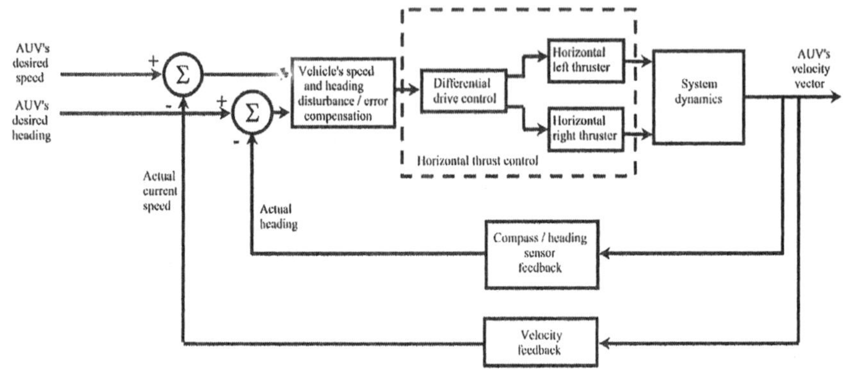

Figure 1. Functional diagram of the trajectory control system.

In this study, we propose a simplified combined model for a particular autonomous underwater vehicle (AUV) developed for entry in an annual intercollegiate design competition conducted by the Association for Unmanned Vehicle Systems International [6]. Since the competition is conducted in an enclosed pool, a simplified model that ignores unpredictable environmental factors is appropriate. The basic configuration of the desired control system for the AUV is identical to a robot platform described by Noland et al.[7]. A functional diagram of the control system modeled is shown in Fig. 1.

The control system for the AUV is modeled using the Arena simulation environment. Validation of the combined discrete-continuous model is performed by comparing the responses with an appropriate controller design in Simulink, based on the derivation of a frequency domain transfer function for the AUV system dynamics.

The rest of this paper is organized as follows: Section 2 describes the AUV system dynamics, and Section 3 discusses the development of the combined discrete-continuous simulation model, while Section 4 describes the validation study of the developed model using results from the Simulink design. Section 5 discusses a supervisory control function for the developed model and finally Section 6 summarizes the results and contributions of the study.

2. AUV SYSTEM DYNAMICS

Since the designed AUV operates in an enclosed pool, several simplifying assumptions could be made. Further development of the model upon this framework, however, could extend its application to an AUV operating in an open water environment. In addition, the focus of this study is on the methodology employed in the development of the model, so a high level of accuracy, i.e. a level of +/-5%, in the determination of the system dynamic parameters was not expected. It was desired, however, to obtain a reasonable approximation of the actual vehicle, which could serve as a guideline and be improved upon as additional data became available. Moreover, it was assumed that environmental disturbances could be ignored. Also since the AUV uses only one pair of thrusters for horizontal motion control, the equation for motion in the direction perpendicular to the AUV frame could be eliminated. It was also decided that the terms representing the cross coupling effects between motions in different planes could be neglected without causing a significant departure from actual AUV behavior. Therefore, the following two differential equations were considered to be sufficient for describing the AUV dynamics [8]:

$$mu' = X_u u' + X_{uu} u\,|u| + U_x \tag{1}$$

$$Ir' = N_r r' + N_{rr} r\,|r| + U_r \tag{2}$$

Equation (1) represents the relation of mass times acceleration equals force. The term U_x is the force in the X direction produced by the thrusters. The X_u and X_{uu} coefficients describe the drag and added-mass effects that oppose the thruster force. Equation (2) represents the relation of moment of inertia times angular acceleration equals torque. The N_r and N_{rr} coefficients describe the drag and added-mass effects that oppose the torque produced by the difference in thruster forces times the moment arm produced by the thruster to frame center dimension.

The mass of the AUV was known to an accuracy of less than one percent. The moment of inertia was estimated by assuming the vehicle mass to be concentrated at four points symmetrically located about the vehicle frame. On the other hand, the determination of drag and added-mass coefficients was done by scaling the coefficient values determined for a similar vehicle described by Yuh [9]. After determining numerical values for the AUV motion parameters, the next step was to characterize the forces generated by the thrusters. The nonlinear response of the thruster to applied voltage presents a problem for the controller design. There are several approaches to the design of control systems that have inherent nonlinear behaviors. One approach is to approximate the nonlinear attributes with linear substitutes and then proceed with the classical techniques. Appendix A shows how the thruster response was linearized about the point of one-half maximum thrust to determine the linearized thruster constant (c_{vL}).

Similarly, the $X_{uu}\cdot u\cdot|u|$ term in Equation (1) presented another linearity issue. This issue was addressed in a manner similar to that used for the thruster response. By assuming an equilibrium condition for Eq. (1) with one-half of the maximum available thrust applied, a midrange velocity value could be determined. A linear-point velocity constant (u_{LP}) was calculated and substituted for $|u|$ in Equation (1). Appendix B shows how the frequency domain transfer function for the AUV velocity in terms of applied thruster voltage was derived. Following the velocity control system model design, the heading control system model design proceeded according to a similar methodology as described in Appendix C.

3. THE COMBINED DISCRETE-CONTINUOUS SIMULATION MODEL IN ARENA

The first step in creating the Arena model of the AUV control system was to implement the simulation of the AUV dynamics. This was accomplished through two basic mechanisms. The first of these two mechanisms is the continuous time structure created by the Continuous, Rates, and Levels elements. This structure produces simulated continuous integration of the Levels variables according to the values of the Rates variables. The details of using this continuous time structure are described by Kelton et al.[10]. The second mechanism involved a simple flow diagram that implemented an Assign module for adjusting the values of the Rates variables at a relatively high constant sample rate.

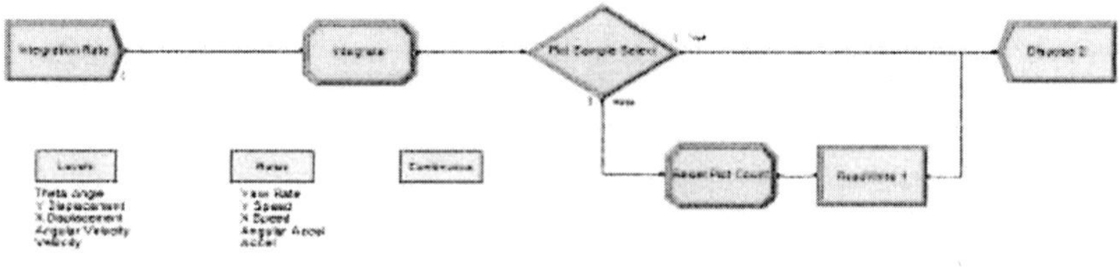

Figure 2. Arena model for AUV dynamics.

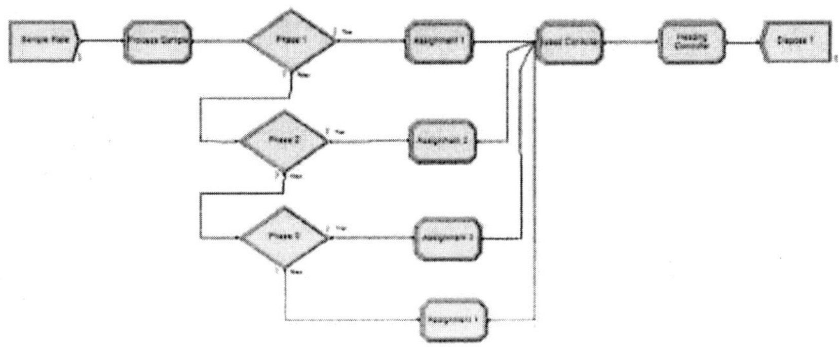

Figure 3. Arena model for AUV controller.

Figure 2 shows a screen print of this portion of the Arena model. The variables listed under the Levels block are integrated continuously according to the corresponding variables listed under the Rates block. The Integration Rate create module produces entities at a constant rate of 100 per second. The Integrate assign module assigns new values to the Rates variables according to certain Levels and other global variables. The decide module and the associated false condition branch serve only to select a proportionally lower sample rate for writing data to a file for external analysis.

Table 1 lists the contents of the Integrate assign module shown in Figure 2. Rows 1 through 16 perform the adjustments to the Rates variables. This is the structure that actually simulates the differential equations (1) and (2). One important aspect of this portion of the model is the selection of either the nonlinear system response or a linear approximation. Setting the global variable Linear Model to a nonzero value selects the linear approximation. The details of how this is accomplished may be understood by examining rows 5, 6, 11, and 16 in Table 2. Rows 17 through 25 of Table 2 deal with higher level functions within the simulation model such as the instrument panel animation and simulated mission navigation.

While the flow diagram of Fig. 2 simulates the "real world" behavior of the AUV based on the physical dynamic laws of motion, the flow diagram shown in Fig. 3 simulates the AUV computational and control capabilities. The remaining blocks of Fig. 3 simulate the AUV trajectory controller. The Sample Rate create module produces entities at a constant rate defined by the reciprocal of the global variable Sample Time. This determines the sample rate for the trajectory control algorithms. Table 2 lists the contents of the Process Sample assign module. Rows 1 and 2 of Table 3 show the pertinent statements that simply assign values to the Speed Sensor and Heading Sensor variables from the instantaneous current values of the "real world" AUV velocity and heading. Rows 3 through 8 of Table 2 deal with the higher level AUV navigation functions and will be described in a subsequent section.

Table 1. Assignments for Integrate assign module.

Row	Variable Name	New Value
1	Yaw Rate	Angular Velocity
2	Heading	AMOD(Theta Angle,6.2832) / 3.1416 * 180
3	Y Speed	Velocity * SIN(Theta Angle)
4	X Speed	Velocity * COS(Theta Angle)
5	PORT(PT) Thrust	PT Control * ((Linear model == 0) * ABS(PT Control) * Thrust Constant + (Linear model <> 0) * Thrust Lin Const)
6	STARBOARD(SB) Thrust	SB Control * ((Linear model == 0) * ABS(SB Control) * Thrust Constant + (Linear model <> 0) * Thrust Lin Const)
7	Avg Accel	(Prev Accel + Accel) / 2
8	Prev Accel	Accel
9	Avg Velocity	(Prev Velocity + Velocity) / 2
10	Prev Velocity	Velocity
11	Accel	(PT Thrust + SB Thrust) / AUV m – (AUV Xu / AUV m) * Avg Accel(AUV Xuu / AUV m) * Avg Velocity * ((Linear model == 0) * ABS(Avg Velocity) + (Linear model <> 0) * Velocity LP)
12	Avg Ang Acc	(Prev Ang Accel + Angular Accel) / 2
13	Prev Ang Acc	Angular Accel
14	Avg Ang Vel	(Prev Ang Vel + Angular Velocity) / 2
15	Prev Ang Vel	Angular Velocity
16	Angular Accel	((SB Thrust – PT Thrust) * Thruster Arm) / AUV I – (AUV Nr / AUV I) *Avg Ang Accel – (AUV Nrr / AUV I) * Avg Ang Vel * ((Linear model == 0) * ABS(Avg Ang Vel) + (Linear model <> 0) * Angular LP)
17	PT FWD	(PT Thrust > 0) * PT Thrust
18	SB FWD	(SB Thrust > 0) * SB Thrust
19	PT AFT	(PT Thrust < 0) * ABS(PT Thrust)
20	SB AFT	(SB Thrust < 0) * ABS(SB Thrust)
21	X Position	X Displacement + X Offset
22	Y Position	Y Displacement + Y Offset
23	WP RANGE X	WAYPOINT X – X Position
24	WP RANGE Y	WAYPOINT Y – Y Position
25	WP RANGE	SQRT(WP RANGE X**2 + WP RANGE Y**2 + 1 / 10**6)
26	WP HEADING	ACOS(WP RANGE X / WP RANGE) * 57.3
27	Plot Count	Plot Count + 1

The remaining blocks of Fig. 3 simulate the AUV trajectory controller. The Sample Rate create module produces entities at a constant rate defined by the reciprocal of the global variable Sample Time. This determines the sample rate for the trajectory control algorithms. Table 2 lists the contents of the Process Sample assign module. Rows 1 and 2 of Table 3 show the pertinent statements that simply assign values to the Speed Sensor and Heading Sensor variables from the instantaneous current values of the "real world" AUV velocity and heading. Rows 3 through 8 of Table 2 deal with the higher level AUV navigation functions and will be described in a subsequent section.

Table 2. Assignments for Process Sample assign module

Row	Variable Name	New Value
1	Speed Sensor	Velocity
2	Heading Sensor	Heading
3	DR POS X	DR POS X + Speed Sensor * COS(Heading Sensor * (3.1416 / 180))* Sample Time
4	DR POS Y	DR POS Y + Speed Sensor * SIN(Heading Sensor * (3.1416 / 180)) * Sample Time
5	DR RANGE X	WAYPOINT X - DR POS X
6	DR RANGE Y	WAYPOINT Y - DR POS Y
7	DR RANGE	SQRT(DR RANGE X**2 + DR RANGE Y**2 + 1 / 10**6)
8	DR HEADING	ACOS(DR RANGE X / DR RANGE) * (180 / 3.1416)

Table 3 lists the contents of the Speed Controller assign module shown in Fig. 3. Rows 1 through 6 implement the proportional, integral, derivative (PID) speed control, which is used in the Simulink model for validation purposes. On the other hand, Table 4 lists the contents of the Heading Controller assign module shown in Fig. 3. As with the speed controller, rows 1 through 6 implement the PID heading control. The statements in rows 9 through 16 perform a procedure for determining the individual thruster voltages for performing a heading change while affecting the velocity as little as possible. The final results of the thruster voltage computation are assigned to another pair of variables in rows 17 and 18. This operation prevents the integration structure from processing intermediate results of the computation and causing improper thruster actuation.

Table 3. Assignments for Speed Controller assign module.

Row	Variable Name	New Value
1	SC Prev E	SC Curr E
2	SC Curr E	SC Set point - Speed Sensor
3	SC P Term	SC P Gain * SC Curr E
4	SC I Term	SC I Term +SC I Gain * (SC Prev E + SC Curr E)/2 * Sample Time
5	SC D Term	SC D Gain *(SC Curr E - SC Prev E)/ Sample Time
6	Thrust Voltage	SC P Term + SC I Term + SC D Term
7	Thrust Voltage	(Thrust Voltage <= Thrust Saturation) * Thrust Voltage + (Thrust Voltage > Thrust Saturation) * Thrust Saturation
8	Thrust Voltage	(Thrust Voltage >= (-1 * Thrust Saturation)) * Thrust Voltage + (Thrust Voltage < (-1 * Thrust Saturation)) * (-1 * Thrust Saturation)

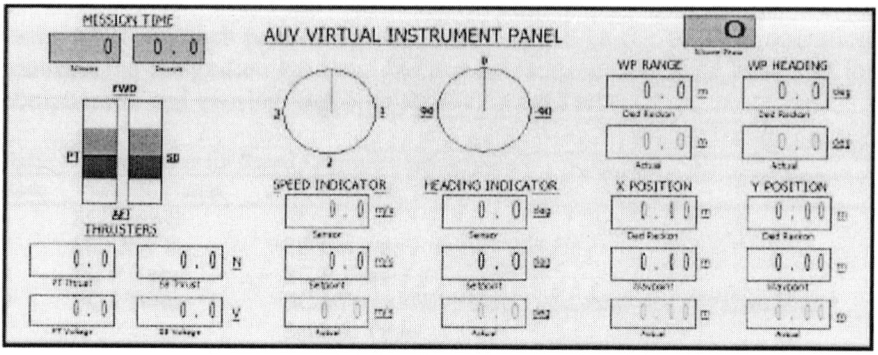

Figure 4. Arena model virtual instrument panel.

Creating a virtual instrument panel for the AUV, as shown in Fig. 4, animated the Arena model. The left hand and center portions of the panel are related to the basic AUV motion control. The left hand area shows the thruster action graphically and numerically. The center portion shows the speed and heading information. The dial gauges each have three different colored pointers that correspond to the colors of the digits on the numeric displays. The right hand portion of the panel concerns the mission navigation control that is discussed in the following section.

Table 4. Assignments for Heading Controller assign module.

Row	Variable Name	New Value
1	HC Prev E	HC Curr E
2	HC Curr E	HC Set point - Heading Sensor
3	HC P Term	HC P Gain * HC Curr E
4	HC I Term	HC I Term +HC I Gain * (HC Prev E + HC Curr E)/2 * Sample Time
5	HC D Term	HC D Gain * (HC Curr E - HC Prev E) / Sample Time
6	Thrust Diff	(HC P Term + HC I Term + HC D Term)
7	Thrust Diff	(Thrust Diff <= (Thrust Saturation * 2)) * Thrust Diff + (Thrust Diff > (Thrust Saturation * 2)) * Thrust Saturation * 2
8	Thrust Diff	(Thrust Diff >= (-2 * Thrust Saturation)) * Thrust Diff + (Thrust Diff < (-2 * Thrust Saturation)) * (-2 * Thrust Saturation)
9	PT Voltage	Thrust Voltage - Thrust Diff / 2
10	SB Voltage	Thrust Voltage + Thrust Diff / 2
11	SB Steering	(PT Voltage > Thrust Saturation) * (PT Voltage - Thrust Saturation)
12	PT Steering	(SB Voltage > Thrust Saturation) * (SB Voltage - Thrust Saturation)

Row	Variable Name	New Value
13	PT Voltage	PT Voltage - PT Steering
14	SB Voltage	SB Voltage - SB Steering
15	PT Voltage	(PT Voltage <= Thrust Saturation) * PT Voltage + (PT Voltage > Thrust Saturation) * Thrust Saturation
16	SB Voltage	(SB Voltage <= Thrust Saturation) * SB Voltage + (SB Voltage > Thrust Saturation) * Thrust Saturation
17	PT Control	PT Voltage
18	SB Control	SB Voltage

4. VALIDATION THROUGH A SIMULINK MODEL

The Simulink model in Fig. 5 contains several elements in addition to the transfer function representing the AUV dynamics. By comparing the model to the diagram of Fig. 1, most of the elements can be identified. The vehicle's speed compensation is composed of a PID controller. Each of the three control terms has an associated gain function and the terms are summed to produce the actuator output signal. The saturation function is inserted between the controller output and the vehicle dynamics to bound the magnitude of the controller output. The controller may reverse the polarity of the output voltage in order to reverse the thrust direction. While it is not intended to model reverse motion of the AUV, the velocity control can be much more responsive by allowing the controller to use reverse thrust for slowing the vehicle.

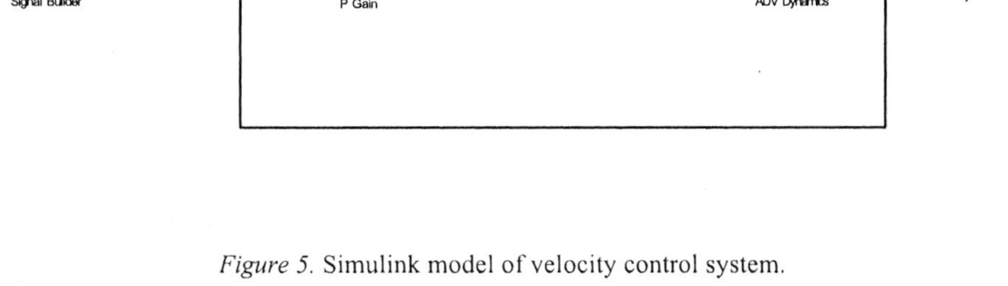

Figure 5. Simulink model of velocity control system.

The input excitation for the Simulink model is provided by the Signal Builder source. A desired speed change for the AUV from the command (or supervisory) control structure would constitute a step input to the controller. Since it was desired to examine the response of the control system for both speed increases and decreases, the Signal Builder was used to produce two successive step functions. The first step is from zero to one at time zero. The second step is from one to zero at a time value of fifteen seconds. A thirty-second observation period is used to compare the system responses to a command to accelerate from 0 to 1m/s followed by a command to stop fifteen seconds later. The output actuating signal for this controller is a voltage signal that creates a corresponding thrust through the two thruster units working together.

The Simulink model of the heading control system is similar to the velocity control system model. A significant difference between the velocity and heading controllers is noted at the end of Appendix C. The task of the heading controller is to produce a desired change in angular displacement, so the controller must produce an angular acceleration of the AUV and then stop the angular motion at the desired heading. Since the angular displacement is the integral of the angular velocity, the angular velocity transfer function is multiplied by the Laplace transform integrator function 1/S. The output actuating signal for this controller is a voltage differential between the input to the two thruster units. The torque for producing an angular acceleration of the AUV results from the difference in thrust between the two thrusters times the moment arm provided by the distance from the center of the thruster to the centerline of the AUV frame. The PID controller functions from the two Simulink models can be easily converted to time domain functions.

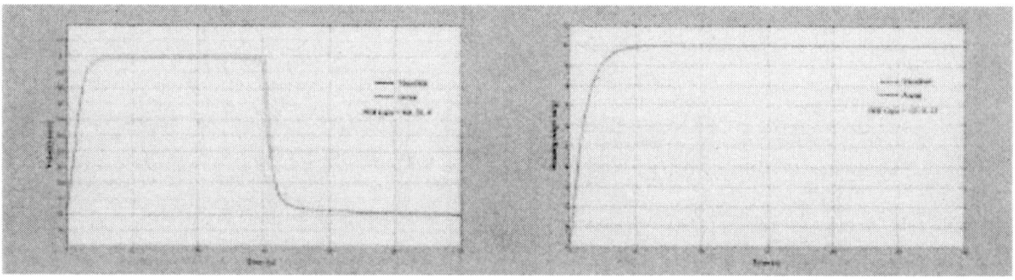

Figure 6. Plots comparing Simulink and Arena linear model velocity and heading step responses.

The responses from the two developed models are used to validate the combined discrete-continuous simulation model. Figure 6 shows plots comparing Simulink and Arena linear model velocity and heading step responses. The close agreement of the two lines demonstrates that the two models produce nearly identical results.

5. ARENA MODEL SUPERVISORY CONTROL

It was desired to use the Arena model to obtain a measure of the performance of the simulated AUV control system. This objective was realized by creating a supervisory control structure to perform the function of navigating the simulated AUV to defined waypoints. This would simulate a possible mission scenario where vision and acoustic sensors were not available, but the mission activity station locations were accurately known. Figure 7 shows the mission navigation diagram taken from the Preliminary Mission Statement for 2006 AUV Competition [11]. Vision subsystems on the AUV platform would normally be employed for locating the Docking Station and the Pipeline Inspection station. A passive acoustic subsystem would normally be used to locate a pinger at the center of the Surface Zone station.

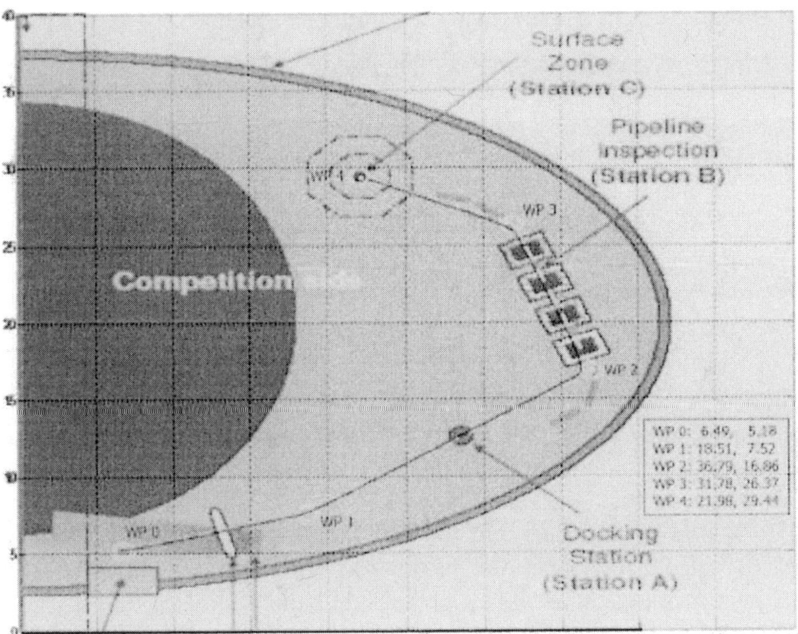

Figure 7. AUV X-Y position track plotted on mission navigation diagram

The waypoint (WP) X and Y coordinates shown on Fig. 7 were determined by pasting the diagram of the competition venue [11] into Microsoft PhotoDraw. The length of the bridge over the pond was estimated to be approximately 40 meters from an aerial photograph of the TRANSDEC facility. The lower left hand corner of the bridge on the diagram provided a convenient origin from which to reference the waypoint coordinates. The cursor position display in PhotoDraw provided the means for constructing accurate waypoint coordinates. A desired track for the AUV was added to the diagram along with perpendicular axis markers from the assumed origin. It must be recognized that this is an artificial view of the competition mission, but it does constitute a reasonable approximate model of the actual situation.

Implementing the navigation functions into the Arena model proved to be an exercise in application of trigonometry. In keeping with the general philosophy of the model, both an actual (or "real world") view and an AUV imperfect view of the navigation variables were created. The actual view is computed at each sample time by the Integrate assign module in the flow structure of Fig. 2. Table 1 rows 21 through 26 show the navigation computations for the actual position of the simulated AUV. The X Offset and Y Offset variables were created to account for the starting position being offset from the origin of the waypoint coordinate system.

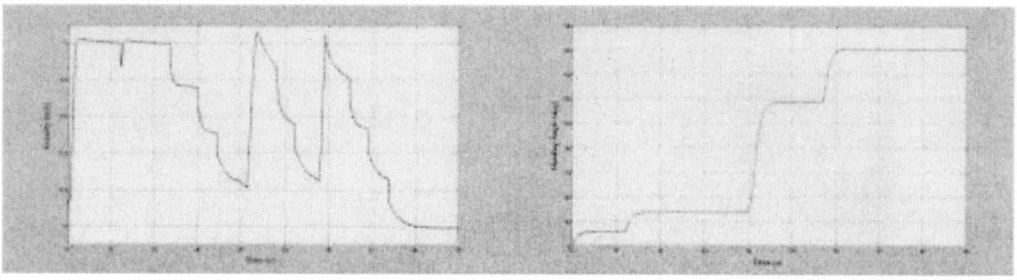

Figure 8. Plots of simulated mission AUV velocity and AUV heading versus time.

The AUV view of the navigation function involves the process of deduced reckoning (DR). The simulated AUV computes the DR navigation variables at each sample time of the Process Sample assign module in the flow structure of Fig. 3. Table 2 rows 3 through 8 show the computation of the DR X and Y positions and the DR range and heading to the current waypoint. All of the computed navigation variables are animated on the virtual instrument panel as shown in the right hand portion of Fig. 4.

The final piece of the navigation capability is the supervisory control decision structure as shown in the central portion of the flow structure of Fig. 2. This structure creates a sequence of mission phases that represent the AUV navigating between pairs of waypoints. The WP 0 coordinates were entered as the initial values for the X Offset and Y Offset variables. The WP 1 coordinates were entered as the initial values for the WAYPOINT X and WAYPOINT Y variables. The condition for advancing to the next mission phase is arrival at a position within one meter of the current waypoint.

The AUV heading set point is updated at each sample time to the computed DR heading to the current waypoint. The speed set point is updated at each sample time based on the range to the waypoint. The idea is for the vehicle to reduce speed in steps as it approaches the waypoint in order to improve the navigation accuracy. Although there is no mechanism for degrading the navigation accuracy in the current model, this is an area of interest for further experimentation with the simulation. Figure 8 shows plots of simulated mission AUV velocity and AUV heading versus time. These plots provide insight into the performance of the simulated controllers under what would be typical operating conditions.

Since the plotting function within the Arena environment only supports plotting a variable against simulation time, it was not possible to animate the trajectory of the AUV. It did prove to be possible, however, to plot the trajectory on the mission navigation diagram as shown in Fig. 7. The plot of Fig. 7 shows that the simulated AUV did follow very close to the desired trajectory. This diagram shows an almost perfect track due to the fact that the controllers were carefully tuned and no sources of error were included in the model. Some interesting plots should be created as various degradation effects are added to the model.

6. CONCLUSIONS AND FUTURE WORK

This study describes the development of a simplified combined discrete-continuous simulation model for an autonomous underwater vehicle. The process involves deriving a differential equation model of the AUV dynamics and designing appropriate velocity and heading control functions. A functioning model was realized in the ARENA simulation environment whereby an animated control panel and output data files demonstrated its successful execution. The resulting data indicates that the model behaves as expected. While a number of simplifications and approximations were involved, the explicit nature of the model structure should facilitate future improvements to the model as better data is obtained from the actual AUV. Furthermore, it was shown that a general purpose simulation package such as Arena could be used to model the system dynamics for an AUV. When compared with other tools used in AUV simulation, the building of the model in Arena has been significantly uncomplicated.

The model can be easily modified to create a mechanism for exploring effects of internal communication delays and sensor errors on the AUV performance, but some amount of additional effort will be required to determine appropriate parameters for the stochastic processes. It has also been recognized that some of the AUV subsystems could possibly be added to the model. An example would be simulating the vision system for locating the lighted docking station. A vision submodel could be created that had some probability of detecting the light as a function of distance and bearing to it. Likewise, a simulation submodel of a passive acoustic navigation subsystem could also be created with random noise added to the ping detection mechanism.

An enhancement that is currently being worked on is the development of a script for Matlab that would process a data file containing a number of different data sets. Simply running the script file could then produce appropriate plots. The ability to very quickly produce and examine the plots would speed the process of experimentation with various scenarios. It may also be possible to construct data sets for analysis by the process optimization utility included with the Arena software package.

APPENIX A: AUV THRUSTER MODEL

From Seabotix data: $\tau_{max} := 2.2\text{kgf}$ $\qquad \tau_{max} = 21.5746\ \text{N}$

$$c_V := \frac{\tau_{max}}{V_{max} \cdot |V_{max}|} \qquad c_V = 0.028\ \frac{s^4 \cdot A^2}{kg \cdot m^3}$$

$V_{max} := 28V$ $\qquad \tau(v) := c_V \cdot v \cdot |v|$

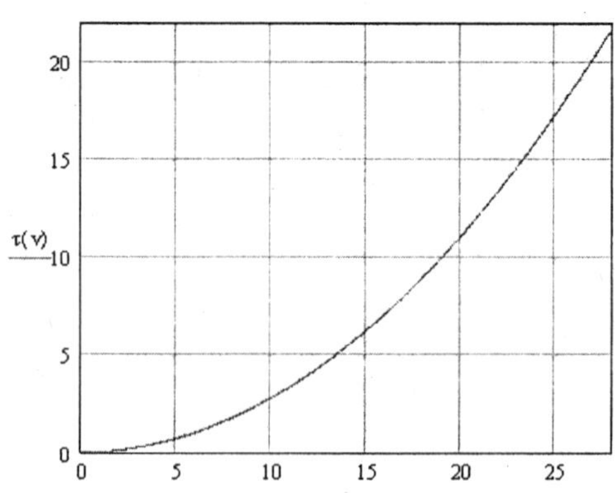

Figure 9. Thruster response curve.

Linearizing the thruster curve at $\tau = \tau_{max} / 2$

$$v_L := \sqrt{\frac{0.5 \cdot \tau_{max}}{c_v}} \qquad v_L = 19.799\ V$$

$$c_{vL} := \frac{0.5\tau_{max}}{v_L} \qquad c_{vL} = 0.545\ \frac{s \cdot A}{m}$$

$$\tau_L(v) := c_{vL} \cdot v$$

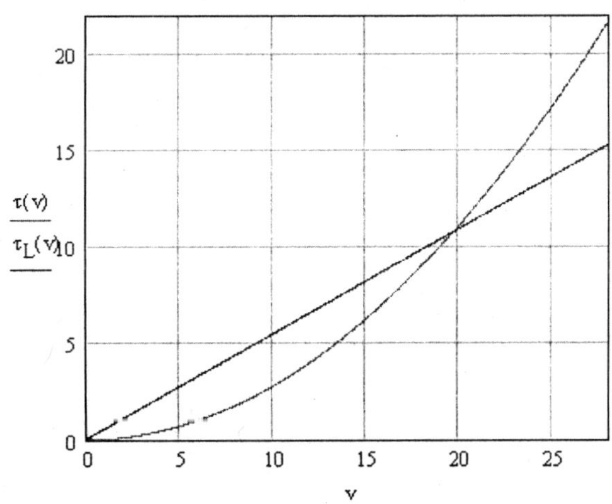

Figure 10. Linearized thruster response curve.

APPENDIX B: DERIVATION OF AUV VELOCITY TRANSFER FUNCTION

Derivation of the transfer function for the AUV velocity in terms of thruster control voltage:

We start with the differential equation (1). When the vehicle is moving at constant velocity, the thruster force is canceled by the hydrodynamic friction forces. (Note that an appropriate sign convention must be applied).

$$0 = 0 + X_{uu} \cdot u \cdot |u| + U_x \tag{3}$$

The squaring of the velocity creates another linearity issue that must be addressed. This equation can be linearized about the velocity value produced by one half of the available thrust.

$$u_{LP} = \sqrt{\frac{-U_X}{X_{uu}}} \qquad u_{LP} = 1.279 \frac{m}{s}$$

The velocity differential equation can now be written as:

$$m \cdot \frac{d}{dt} u(t) = -X_u \cdot \frac{d}{dt} u(t) - X_{uu} \cdot u_{LP} \cdot u(t) + 2 \cdot c_{vL} \cdot v(t) \tag{4}$$

where the applied force U_X has been replaced by the output of the two thruster in terms of the applied voltage. Taking the Laplace transform gives:

$$m \cdot s \cdot U(s) = -X_u \cdot s \cdot U(s) - X_{uu} \cdot u_{LP} \cdot U(s) + 2c_{vL} \cdot V(s) \tag{5}$$

where U(s) is the AUV velocity and V(s) is the thruster control voltage in the *s* plane.

$$\frac{U(s)}{V(s)} = \frac{\text{Velocity}(s)}{\text{Voltage}(s)} = G(s) \tag{6}$$

$$G(s) = \frac{2c_{vL}}{\left(m + X_u\right) \cdot s + X_{uu} \cdot u_{LP}} \tag{7}$$

Substituting the constant values and simplifying gives:

$$G(s) = \frac{1}{30.74 \cdot s + 15.50} \tag{8}$$

APPENDIX C: DERIVATION OF AUV HEADING TRANSFER FUNCTION

Derivation of the transfer function for the AUV angular velocity in terms of thruster control voltage:

When the vehicle is moving with constant angular velocity, the hydrodynamic friction forces cancel the torque produced by the differential thruster force. (Note that an appropriate sign convention must be applied).

$$0 = 0 + N_{rr} \cdot r \cdot |r| + U_{I} \tag{9}$$

The squaring of the angular velocity creates another linearity issue that must be addressed. This equation can be linearized about the angular velocity value produced by one half of the available differential thrust.

The moment arm $A_m := 0.2m$

$$T_{max} := \tau_{max} \cdot 2A_m \qquad T_{0.5} := 0.5 \cdot T_{max}$$

$$\frac{T_{0.5}}{N \cdot m} = 4.315$$

$$r_{LP} = \sqrt{\frac{T_{0.5}}{N_{rr}}} \qquad r_{LP} = 0.61 \cdot \frac{m}{s}$$

The angular velocity differential equation can now be written as:

$$I \cdot \frac{d}{dt} r(t) = -N_r \cdot \frac{d}{dt} r(t) \cdot N_{rr} \cdot r_{LP} \cdot r(t) + 2A_m \cdot c_{vL} \cdot v(t) \tag{10}$$

where v(t) represents the voltage differential between the two thrusters. Taking the Laplace transform gives:

$$I \cdot s \cdot R(s) = -N_r \cdot s \cdot R(s) - N_{rr} \cdot r_{LP} \cdot R(s) + 2A_m c_{vL} \cdot V(s) \tag{11}$$

where R(s) is the AUV angular velocity and V(s) is the differential thruster control voltage in the *s* plane.

$$\frac{R(s)}{V(s)} = \frac{\text{Angular\_Velocity}(s)}{\text{Diff\_Voltage}(s)} = G(s) \quad (12)$$

$$G(s) = \frac{2A_m c_{vL}}{(I + N_r)\cdot s + N_{rr}\cdot r_{LP}} \quad (13)$$

Substituting the constant values and simplifying gives:

$$G(s) = \frac{1}{74.33\cdot s + 32.47} \quad (14)$$

Recognizing that the control system is actually concerned with the angular displacement, this transfer function is multiplied by an integrator in the form of 1/s:

$$G_\theta(s) = \frac{1}{74.33\cdot s^2 + 32.47s} \quad (15)$$

REFERENCES

1. X. Chen, D. Marco, S. Smith, E. An, K. Ganesan, T. Healey, 6 DOF Nonlinear AUV Simulation Toolbox, *Proceedings. of MTS/IEEE OCEANS '97 Conference*, p. 1070.
2. H. Lin, D. Marco, E. An, K. Ganesan, S. Smith, T. Healey, Modeling and Simulation for the FAU AUVs : Ocean Explorer, *Proceedings of MTS/IEEE OCEANS '98 Conference* **3**, p. 1728.
3. F. Song, A. Folleco, E. An, High Fidelity Hardware-In-the-Loop Simulation Development for an Autonomous Underwater Vehicle, *Proceedings of MTS/IEEE OCEANS'01 Conference* **1**, p. 444.
4. F. Song, E. An, A. Folleco, Modeling and Simulation of Autonomous Underwater Vehicles: Design and Implementation, *IEEE Journal of Oceanic Engineering* **28**(2), 2003, p.283
5. NPS Autonomous Underwater Vehicle (AUV) Workbench; http://terra.cs.nps.navy.mil/AUV/workbench/
6. Association for Unmanned Vehicle Systems International; http://www.auvsi.org/

7. S. Noland, L. Molnar, C. Flanagan, Implementing a Layered Control on an Open Framed AUV, *Proceedings of the IFAC Workshop on Guidance and Control of Underwater Vehicles*, 2003, p. 61.
8. D. Yoerger, J. Slotine, Robust Trajectory Control of Underwater Vehicles, *IEEE Journal of Oceanic Engineering* **10**(4), 1985, p. 462.
9. J. Yuh, Modeling and Control of Underwater Robotic Vehicles, *IEEE Transactions on Systems, Man, and Cybernetics* **20**(6), 1990, p. 1475.
10. D. Kelton, R. Sadowski, D. Sturrock, *Simulation with Arena* (McGraw Hill, New York, 2004)
11. Preliminary Mission Statement for 2006 AUV Competition; http://www.auvsi.org/competitions/PreliminaryMissionStatementfor2006AUVCompetition.pdf

EX-POST INTERNET CHARGING: AN EFFECTIVE BANDWIDTH MODEL

Joseph P. Bailey, Ioannis Gamvros, and S. Raghavan
The Robert H. Smith School of Business
University of Maryland
College Park, MD 20742-1815

Abstract Generally Internet Service Providers (ISPs) have charged their customers flat fees for their Internet connections. This has resulted in frequent congestion for many users. There are many different approaches to address this problem. Effective utilization of scarce resources is important to managers in the telecommunications industry, and thus usage-based pricing has become an important tool to address this problem—since it does not require large capital expenditures. In this paper we develop an ex-post charging mechanism based on the effective bandwidth concept. This model, effectively characterizes the utilization and burstiness of a user in a single metric. Further, we introduce a novel market for buffer size. In this market users purchase a specific buffer size from their ISP. Our model directs users with bursty traffic to purchase larger buffers, while users with well-behaved traffic are directed to purchase smaller buffers. From a resource usage standpoint, this is also the appropriate decision. We conduct computational experiments to show the viability of this approach, and also discuss real-world implementation issues.

1. Introduction

Over the past ten years there has been an ongoing debate over the issue of charging Internet traffic (see McKnight and Bailey, 1997). The growing numbers of Internet users coupled with the development of new applications that require large amounts of bandwidth has led to an explosive growth in Internet traffic resulting in frequent congestion that is widely perceived as poor service. More users are growing frustrated by slow connections and increasing packet delays (that result in slow applications like web browsing, ftp, e-mail etc.). Internet Service Providers (ISPs) are trying to solve this problem by over-provisioning (i.e., placing extra bandwidth) in the core of their backbone networks in order to alleviate the congestion experienced. However, there is a growing view amongst a group of researchers that this is a short-term (patch-up) solution that will not solve the problem. These researchers blame instead the charging mechanisms that prevail in the Internet and insist that the Internet congestion problems can be alleviated to a large extent by using more sophisticated charging algorithms instead of investing heavily in faster routers and extra capacity. Investing in capacity is a significant capital expense which is difficult for telecommunications companies in the current market environment, and so it is critical for ISPs to develop pricing schemes to address this problem.

Although there has always been hope that increasing bandwidth availability will alleviate any need for bandwidth charging, problems of Internet congestion appear to be chronic. The supply and demand for Internet bandwidth appear to be in a virtual cycle whereby increasing supply of bandwidth allows for greater use of bandwidth-intensive applications. Increasing use of bandwidth-intensive applications leads to more demand of bandwidth. For example, the use of Internet Protocol networks to deliver video content is just the latest bandwidth-intensive application. As of 2006, there are a large number of users willing to watch compressed video that would be roughly equivalent to over-the-air broadcast quality. Already there are providers like www.movielink.com and www.cinemanow.com that deliver video movies over the Internet. It is reasonable to assume that demand for HDTV or HD-DVD quality video delivered over the Internet is not far behind! In this way, the evolution of the supply and demand of Internet bandwidth is similar to the evolution of memory and software. No matter how much memory is available today, one could imagine that future applications will inevitably demand even more. Recent scholarly articles continue to point to the fact that Internet congestion is chronic (see Srikant, 2004; Low et al., 2002) even though some argue it is not (Odlyzko, 2003). To help support the claim that Internet congestion is chronic, the most recent statistics that gauge Internet congestion (as measured by packet loss or packet delay, for example), continue to show problems of Internet congestion (see IHR, 2006; ITR, 2006).

In this paper we consider organizational users like small businesses, universities, government organizations etc., that lease a connection for their organization to communicate with the Internet. We believe that this is the right set of users to focus on, since the bulk of the traffic when the network is congested (i.e., during day time on weekdays) comes from these users. At present most of these Internet users are charged a price that is dependent solely on their connection bandwidth. In other words users pay a flat-fee every month to their ISP irrespective of the volume of traffic that they send over their connection. Some researchers like Odlyzko, 2001 are in favor of the status quo because they believe that the simplicity of the flat-fee model is essential and that over-provisioning can be more than a short-term solution. Another option is to charge users based on the actual traffic sent over their connection (i.e., usage-based charging). While another viewpoint by Shenker et al., 1996 is that flat-fee and usage-based charging can co-exist in the same market in the same way as they do in telephony. Those that are against flat-fee pricing argue that it leads to the "tragedy of the commons" and should be replaced by a smarter charging mechanism. Right now all Internet users (obtaining service from the same ISP) pay the same price for the same connection speed even though their utilization rates of the connection can vary significantly. As a result low-end users end up subsidizing the high-end or heavy users who are possibly willing to pay more for their service.

Proponents of usage based pricing have already done a significant amount of work on the issue of usage based Internet charging (see MacKie-Mason and Varian, 1995; Falkner et al., 1999; Kelly, 1997; Courcoubetis et al., 1998; Courcoubetis et al., 2000). Most of this research has focused on devising optimal pricing strategies that aim at maximizing social welfare. This is achieved by estimating the marginal congestion created when a user sends a packet. The price the users are charged is proportional to the additional congestion generated by the packets they are sending. One of the most notable approaches that uses the marginal congestion paradigm is proposed in MacKie-Mason and Varian, 1995 where a so-called "smart market" mechanism is explained. In this charging scheme packets are assigned bids that are used to determine which packets are given priority. Usually these optimal approaches suffer from high complexity and difficult, if not impossible, implementations that make them unattractive for the real world. The idea of charging based on marginal congestion costs has also been criticized by Shenker et al., 1996. They claim that 1) marginal cost prices may not produce sufficient revenue to fully recover costs, 2) congestion costs are hard to compute, and 3) there are other structural goals of pricing that marginal congestion cost models do not address.

A different way to solve the congestion problem altogether is to make sure that any user that is given access to the network will under no circumstances slow down the traffic of other users. This is achieved by what is known as Call Admission Control (or simply Admission Control) and it involves policing who is connected to the network, what kind of traffic they are sending and either approving or rejecting more connections from other users. In CAC each user makes a request to the network specifying the traffic characteristics (i.e. peak rate, packet loss and acceptable delays) of the data flow he wishes to send. An admission control algorithm then checks the current status of the network to make sure that there are available resources to support the specific data flow with the required Quality of Service guarantees and either admits the user and assigns a charge to the connection or denies admission. If the connection is admitted then the network is required to monitor the traffic that the user is sending to make sure that it complies with the request that was made originally. One of the approaches that uses CAC is Falkner et al., 1999. In both the CAC and the "smart market" approach the charging mechanism is required to know in advance (in the case of CAC) or follow the entire path (in the case of the smart market approach) that packets take from source to destination in order to assign charges. This may be quite difficult when traffic travels over multiple domains (service providers) to get from source to destination. Consequently, these requirements induce significant overhead and can cause scalability problems.

Unfortunately, proponents of new pricing models and complex admission control policies may never be able to adequately solve Internet congestion. One roadblock is the inability for the Internet to move away from the End-to-End design principles to a "Brave New World" (Blumenthal and Clark, 2001) where competing ISPs can coordinate their activities. Specifically, if ISPs wanted to develop a new "smart market" class of pricing, they would have to develop some settlement process whereby one ISP would reimburse other ISPs for carrying priority traffic. Alternatively, admissions control policies would also have to be closely coordinated in order to implement many QoS solutions. If the market were moving towards more industry concentration, then coordination across so many ISPs would not be a problem. However, ISP backbones have been unsuccessful in their attempts to integrate, in part because of merger guidelines that appear to be too stringent on defining market power (Besen et al., 2002). When there are multiple networks responsible for the transmission of packets, it is difficult to implement an End-to-End pricing scheme for a few reasons. First, it would require all of the involved networks to adhere to the same policy. This is very difficult because these networks are not only competing for the same customers, but they have an incentive to provide better service to their customers in preference over their competitor's customers. Second, an End-to-End pricing scheme may be ripe for opportunism whereby an ISP can try to enhance its settlement money. For example, they may support larger routing tables to carry more traffic and it may even send the traffic over a greater number of hops within its network to increase its portion of the settlement. Therefore, some of the most promising solutions to Internet congestion are the ones that embrace, rather than abandon, the End-to-End design principles of the Internet.

Internet congestion may be reduced by a class of charging mechanisms that assign prices based only on information collected at the ingress of the network, where the user's packets enter. This paradigm is termed "edge pricing" (see Shenker et al., 1996) and it works by monitoring the packets that users send over their connection either constantly or at given intervals. While monitoring, the charging algorithms determine the traffic characteristics of different users and in return are able to estimate the network resources utilized by these users and the congestion they impose on others. Based on this information, charges that are proportional to the resource usage of each user are assigned. Edge pricing does not entail the risks and difficulties of the CAC or smart market approaches, but imposes the challenge of estimating resource consumption based on local information at the ingress point of the network. In many cases this challenge is met with the use of effective bandwidth bounds (Kelly, 1997; Siris et al., 1999) that give good estimates of a users actual resource usage of the ingress connection. In these charging mechanisms users declare a utilization rate at which they will send data over their connections. If they respect this rate then they are charged according to the estimated effective bandwidth. However if their actual rate is different (even if it is lower) from the stated then they get penalized by paying more than what the effective bandwidth calculation indicates.

In this paper we develop a novel model for charging Internet connections based on effective bandwidth. This model falls under the class of the so-called "ex-post charging" models (see Bailey et al., 2006) where the pricing algorithm is determined ex-ante but the charges are determined after the traffic has been sent. Our effective bandwidth model is quite simple, and differs from other effective bandwidth models in the literature in several respects. First, we use the large buffer asymptotic method for calculating effective bandwidth (Guérin et al., 1991). As a consequence, unlike other effective bandwidth models used for pricing, we do not need to consider other sources of traffic to determine the charge for the traffic. This might seem a disadvantage of the model at first, since any possible multiplexing gains are not calculated, but in fact is an *extremely desirable property*. This is because (i) the charge is dependent solely on an individual user's traffic, and (ii) it can be calculated by the user without knowing any other users traffic (and thus can manage their traffic and charges without worrying about the affect of the behavior of other users on their charge). Second, one of the parameters that the effective bandwidth depends upon is a buffer size. We develop a market for buffers where ISPs charge users for buffer space, and based on this develop a coherent pricing model. Finally, our model satisfies a desirable feature that the ex-post charging mechanism has—namely Bayesian updating of parameters. This means terrabytes of traffic information need not be stored to determine the charge for the traffic. This is an important and critical issue, that seems to have been largely ignored in the literature. By ensuring that terrabytes of data need not be stored to implement the pricing mechanism (i) it is more likely to be accepted (ii) cheaply implemented, and (iii) removes a potential security risk associated with storing trace data.

In the rest of this paper we develop our effective bandwidth based pricing model. The remaining sections are organized as follows. In the rest of this section we review the ex-post charging model and philosophy. In Section 2 we will review the large buffer asymptotic model for effective bandwidth and the upper bound based on it that we use in our model. In Section 3 we develop our effective bandwidth charging model, introduce a market for buffers, and discuss issues concerning the fine tuning of the model to the needs of different ISPs. In Section 4 we present numerical results that showcase the performance of our pricing algorithm under different scenarios, and illustrate the behavior of the pricing algorithm. Finally in Section 5 we present our conclusions and suggestions for future work on this area.

1.1 The Ex-Post Charging Approach

As we stated at the outset, ex-post charging may be most suitable between an ISP and organizational customers such as small businesses, universities, and government organizations. These users currently shape or manage their traffic, and are most concerned about their quality of service and of lowering their large Internet connectivity bills. These users are likely to benefit on both counts from the ex-post charging policy we propose and embrace it. On the other hand mass-market users (like residential customers) currently appear to prefer flat rate pricing (as evidenced by the shift in pricing schemes in the mobile and long-distance market). Interestingly, it is precisely for these customers with small individual revenues (in the \$20-\$50 range) the cost of calculating and metering usage makes usage-based pricing a costly proposition for the ISP.

The ex-post charging mechanism falls under the category of "edge-pricing" algorithms. In this model the charging algorithm is determined in advance while the actual charge is calculated after the fact. We note that the actual charging mechanism that might be used in practice will in fact consist of an ex-ante charge (i.e., a charge determined in advance) as well. So in essence the final price P can be viewed as:

$$P = P_{\text{ex-ante}} + P_{\text{ex-post}} \tag{1}$$

The ex-ante part of the price can be used as a mechanism that will prevent users from reserving connections that they don't really need. If there was no such component and the price depended only on resource usage a customer would be able to ask for multiple connections, not send any traffic over them, and pay nothing. Although, we do not study the ex-ante part of the price in this paper it plays an important role as well.[1] For example, it may affect the ability of an ISP to attract new customers in a competitive setting. However, when considering that an ISP has already made a contract with a customer, the ex-ante portion of the charge is sunk and will not affect a customer's incentive to manage its Internet traffic any differently. As we will not consider the ex-ante portion of the price within this paper, from now on we will use the term price to refer to the ex-post price.

For any new charging model to be accepted, and successfully implemented, in the current ISP marketplace, we believe there are two key desirable features—

[1] For example, the ex-ante portion of the price may typically cover the cost of maintaining a connection. Users with large bandwidths may thus have a higher ex-ante portion of the price.

simplicity, and Bayesian updating of parameters—as described above. Additionally, Bailey et al., 2006 identify the following desirable qualitative characteristics of an ex-post Internet charging model.

- The ex-post charge should be a monotonically increasing function of the total volume of traffic sent (and/or received). Utilization measures the volume of traffic divided by the speed of the connection times the duration (over which the volume of traffic is sent). Consequently, the ex-post charge should be a monotonically increasing function of utilization. Further, since it is likely that the provider would probably want to offer a service in which economies of scale are realized it is desirable for the relationship between the ex-post price as a function of the measured utilization to be concave.

- Burstiness expresses the notion of sudden, unpredictable and usually large transmissions of data from the customers to the provider. Bursty traffic can be problematic for an ISP because an ISP must either size their network large enough to accommodate peak periods or be willing to endure periods of congestion during peak periods. Consequently, bursty traffic should be charged a higher price than well-behaved traffic (i.e., if two traces have the same utilization but one is burstier than the other it should be charged more). Additionally, the relationship between the price and the measured burstiness should be a convex function. This corresponds to the notion that the effects of bursty traffic on a provider's network can have an additive effect resulting in prolonged network congestion as customers send more bursty traffic. We note however that burstiness is not a well defined metric like utilization. Consequently, it may not be easy to analytically verify whether a pricing model complies with this desired feature.

- Finally, the implementation of the charging mechanism should be transparent to the network. By transparent we mean that the algorithm should require very few or absolutely no network resources (e.g. bandwidth, CPU time, storage space) to complete its task.

2. Theoretical Background for Effective Bandwidth

In order to be able to charge customers for the use of a communications link we need to be able to identify scalars that will measure the resources they use when their packets are forwarded over the Internet. These scalars will then become the independent variables of a pricing function that will associate resource usage with a specific charge. Utilization of a traffic stream is a well defined metric and is easy to measure. However, burstiness is not so well defined. The effective bandwidth concept ties these two notions together, and summarizes

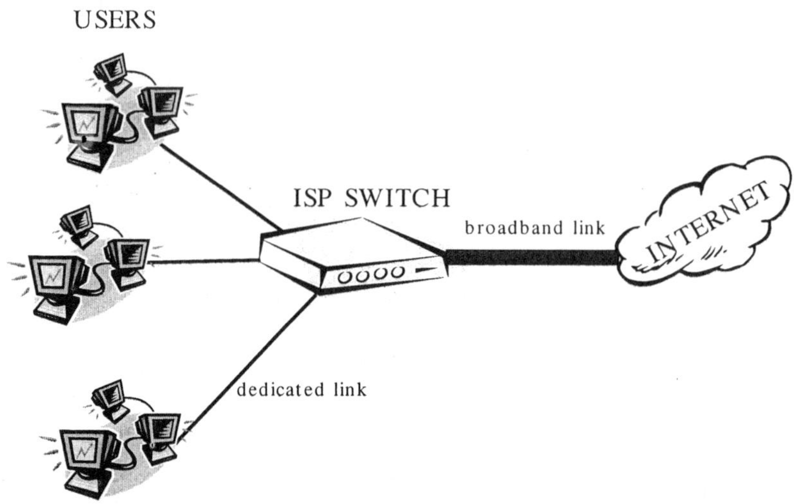

Figure 1. Multiplexing of many sources on an outgoing broadband link.

resource usage of a shared communications link by a specific source. We now review the effective bandwidth concept, and in particular the large buffer asymptotic method for computing effective bandwidth proposed by Guérin et al., 1991.

Effective Bandwidth is a scalar that summarizes resource usage on a communications link from a specific source in a packet-switched network. Specifically, at a given switch in the network where many traffic streams from different sources are multiplexed on a single outgoing link (Figure 1), the effective bandwidth of a specific source represents the capacity of the outgoing link used by that source.

It turns out that the effective bandwidth of a specific source depends not only on the statistical properties of the traffic source in question but also on the statistical properties of the other sources that it is multiplexed with, the characteristics of the switch (i.e., buffer size) and the characteristics of the link that the source utilizes (i.e., capacity). Moreover, effective bandwidth depends on the Quality of Service requirements (i.e., packet loss probability), which are imposed by the source.

A well known and widely accepted way to calculate the effective bandwidth of a specific traffic source is proposed by Kelly, 1996. The calculation proposed there takes into account all of the parameters that we mentioned previously and provides a very good estimate of network resources used by each source. However, Kelly's calculation requires the collection of large amounts of data that represent the traffic each source sends to the switch. This is somewhat impractical from an individual users perspective, since to determine their charge for the connection they would need to know the traffic of other users. As a result instead of Kelly's model we will use an upper bound for effective bandwidth that was proposed by Guérin et al., 1991 and is easy to calculate.

Guérin et al. use what is called the large buffer asymptotic method to arrive at an *upper bound* for the effective bandwidth of a single source. This method is concerned with the overflow probability of the buffer at the switch as the buffer size increases. Additionally Guérin et al. do not take into account the traffic characteristics of other sources that send data to the switch. At first this seems to be a significant disadvantage of the calculation as it completely ignores possible gains in resource usage from multiplexing. However, for the purposes of pricing this model is *ideal* since the scalar that represents the resource usage for a specific user depends solely on the user's traffic and not the combined traffic characteristics of all the users that send traffic to the network. In simpler terms each customer is charged based solely on their individual traffic characteristics without considering traffic characteristics of other users. Since we are interested in using the effective bandwidth concept for pricing and not for traffic engineering the upper bound on the effective bandwidth is adequate as a measure of resource usage for our purposes.

2.1 Effective Bandwidth of a Single Source

In the following we review Guérin et al.'s large buffer asymptotic model for effective bandwidth (see Guérin et al., 1991). They assume that the traffic sources can be in one of two possible states at a time. Either the source is in a "Burst State" which means that it is sending data at the maximum rate of the connection or it is in an "Idle State" which means that there is no transmission. This assumption actually states what is happening on communication links that are utilized on the Internet. In order to be able to fully characterize the traffic source one needs to know the distributions of the "Burst State" periods and the "Idle State" periods. The second assumption that they make states that the length of the "Burst State" and "Idle State" periods are exponentially distributed. As a result they can be completely characterized by their means (i.e., mean length of the "Burst State" and "Idle State" periods).

Consequently, a traffic source can be fully identified if we know the peak rate (R_p) at which it can transmit, the mean of the "Burst State" periods (b) and the mean of the "Idle State" periods. Observe that given the mean of the "Burst State" periods and the mean of the "Idle State" periods one may calculate the utilization (ρ) as the mean of the "Burst State" periods divided by the sum of the means of the "Burst State" and "Idle State" periods. As a result given the source's utilization (ρ) and mean of the "Burst State" periods the mean of the "Idle State" periods can be computed. Therefore a traffic source can be fully identified if we know the peak rate (R_p), the mean of the "Burst State" periods (b), and the source's utilization (ρ).

We are interested in calculating the capacity (C) in bits per second (bps) that for a specific buffer size (B) guarantees a buffer overflow probability less than ϵ. The capacity (C) is the effective bandwidth of the source and it means that the outgoing link shown in Figure 1 should have at least C bps reserved for the traffic source in question in order to be able to support the source's traffic for a specific overflow probability and the given buffer size. Guérin et al. show that an upper bound on C is given by the following equation:

$$C = \frac{\gamma b(1-\rho)R_p - B + \sqrt{(\gamma b(1-\rho)R_p - B)^2 + 4B\gamma b\rho(1-\rho)R_p}}{2\gamma b(1-\rho)} \quad (2)$$

where $\gamma = \ln(1/\epsilon)$.

This equation provides us with an estimate of the actual effective bandwidth of the source. Numerical experiments have shown that the value of C calculated by Equation 2 is very close to the exact value of effective bandwidth (see Guérin et al., 1991). With the help of Equation 2 we have a good approximation of the resource usage of a specific source and in turn of a specific customer.

2.2 Implementation Issues

We now show that the variables that are needed for the calculation of the effective bandwidth are readily available or can be easily measured. Specifically, the peak rate R_p of the connection is known in advance, the buffer size B that resides in the switch where the link that carries the customer's traffic is connected can be easily verified and the packet loss probability that the customer requests is agreed upon in the service level agreement. That leaves us with the actual measurement of the mean burst period b and the utilization ρ.

In order to calculate these values one needs to know the size of the packets that the source has transmitted and the time of their arrival at the port of the switch. Once the arrival times and the sizes of the packets are known, the mean burst period and the utilization can be calculated as follows. Utilization is given by the sum of the packet sizes (total volume of traffic) divided by the connection speed times the total period of measurement (i.e., divided by the maximum possible traffic that could have been sent on the connection over the duration of measurement). In order to calculate the mean burst period we have to determine consecutive packets that went over the connection in a burst (i.e., these packets were sent one after the other with no idle time between them). We define a series of sequential packets arriving over the span of a millisecond to be in the same burst. This assumption is supported by the fact that most measuring equipment cannot discriminate between arrival times under the millisecond level (see Mills, 1989).[2] We set the size of the burst equal to the sum of the sizes of all the packets in the burst. Also we set the arrival time of the burst equal to the arrival time of the first packet in the sequence. The mean burst period of the entire data trace can then be calculated by first calculating the mean size of a burst as the sum of the sizes of the bursts (which is equal to the total volume of traffic) divided by the number of the bursts; and then dividing the mean size of a burst by the connection speed.

It is important to point out here that there is no need to store large data files that contain the above information. For every new packet arrival the mean of the burst period and the utilization can be updated (since the mean of a time series can be updated in a Bayesian fashion) resulting in a new value for effective bandwidth. As a result the storage requirements for the calculation are minimal and the resource measurement can be done in a meter like fashion in a similar way as with utilities such as electricity and natural gas.

3. The Effective Bandwidth Ex-Post Charging Model

We now build an ex-post charging model based on effective bandwidth.

3.1 The Simple Effective Bandwidth Model

Since effective bandwidth is a good measure of resource usage it makes sense that a charging model could in fact consist of just the value of the effective bandwidth and not take any other variables into consideration. A possible model would look something like this:

$$P = a * C \tag{3}$$

where P is the ex-post price, C is the value of the effective bandwidth calculated in Equation 2 and a is a variable that can change during the billing period and its purpose is explained below.

The basic characteristic of this charging model is its simplicity. The price changes linearly with effective bandwidth and as a result it directly corresponds to a user's resource usage. The variable a is used to convert C into monetary units and its units are dollars per bps per unit of time. In addition a reflects congestion in the network and contention among the users for the scarce resources. As a result when the demand is high the ISP can increase the value of a whereas at times of the day when the demand is low the ISP can accordingly reduce a in order to attract more users. We note further that the model can be used to calculate the price on whatever time interval basis the ISP and/or user agree upon. For example, the effective bandwidth may be computed on an hourly

[2] It is possible to get measurements that are accurate up to a microsecond level (see Micheel et al., 2001) but that requires more sophisticated techniques and equipment.

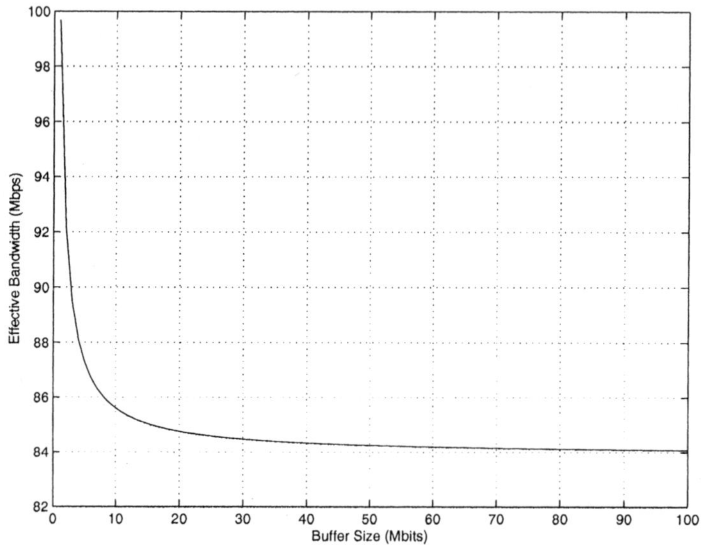

Figure 2. The effect of buffer size on effective bandwidth

basis (for the traffic sent to (and/or received) the ISP in the past hour) and the total charge for that hour may be determined using Equation 3. Although the value that a assumes is extremely important it is beyond the scope of this paper and we will not discuss it any further.

There is only one real disadvantage with the simple effective bandwidth model. The problem is that the value of effective bandwidth and consequently the price that customers pay for the service depends on the buffer size that is reserved for them. Everything else being equal (i.e., utilization, mean burst period, peak rate) the effective bandwidth value decreases with an increase in buffer size. This is shown in Figure 2 for a 90 second trace captured on an OC-3 link.

Looking at Figure 2 (and Equation 2) it should be clear that the choice of buffer can significantly affect the calculation of effective bandwidth and the price calculated by the model. As a result the customer will always want to have as large a buffer as possible while the ISP would prefer *exactly the opposite* since smaller buffers would mean higher prices and higher revenues. Additionally, buffer size is an important resource that can play a critical role in terms of the quality of service that the customer receives. Customers with large buffers can send large bursts of data to the ISP and be sure that their packets are not going to be dropped. As a result, we believe, buffer size should be taken into account in any charging mechanism for any packet switched network. Further,

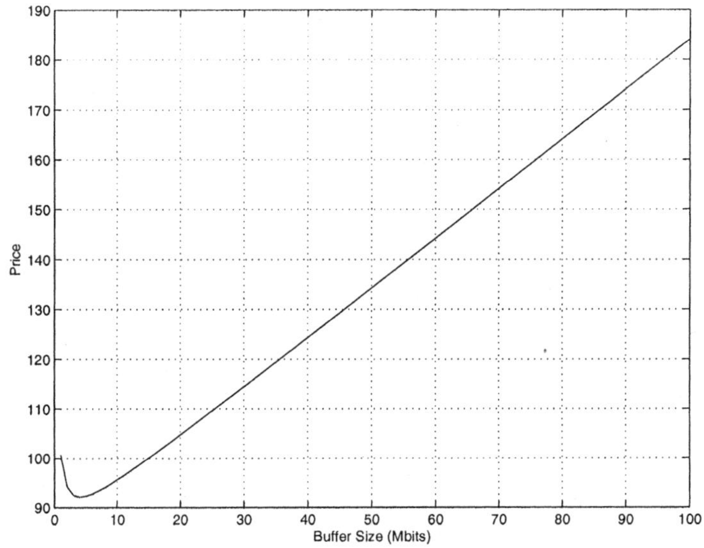

Figure 3. The calculated charge is overwhelmed by the choice of the buffer size

buffer is a resource provided by the ISP, and thus a cost to the ISP, and so we believe that it should play a role in the pricing model.

The next charging model that we propose resolves the shortcomings of the simple effective bandwidth model as it takes into account the buffer size that is reserved for the customer's traffic as well as the effective bandwidth calculated for that buffer size.

3.2 The Delta Model

In this model we assume that there is a market for buffer sizes and the customers are able to select buffer sizes that are consistent with their traffic demands. For example the model should direct customers with bursty traffic to choose larger buffers while customers with well-behaved traffic should be rewarded for selecting a smaller buffer. The Delta charging model that we propose is based on these assumptions and has the following form:

$$P = a * (\Delta * B + C) \tag{4}$$

where P is the ex-post price component in dollars per unit of time, C is the effective bandwidth in bps, B is the buffer size in bits and a is a variable that has the same role as in the previous model.

Delta (Δ) is a scaling constant that has as a primary purpose to balance the effect of the two resources B and C on the price. This is done because the buffer size can assume values that are comparable and some times even greater

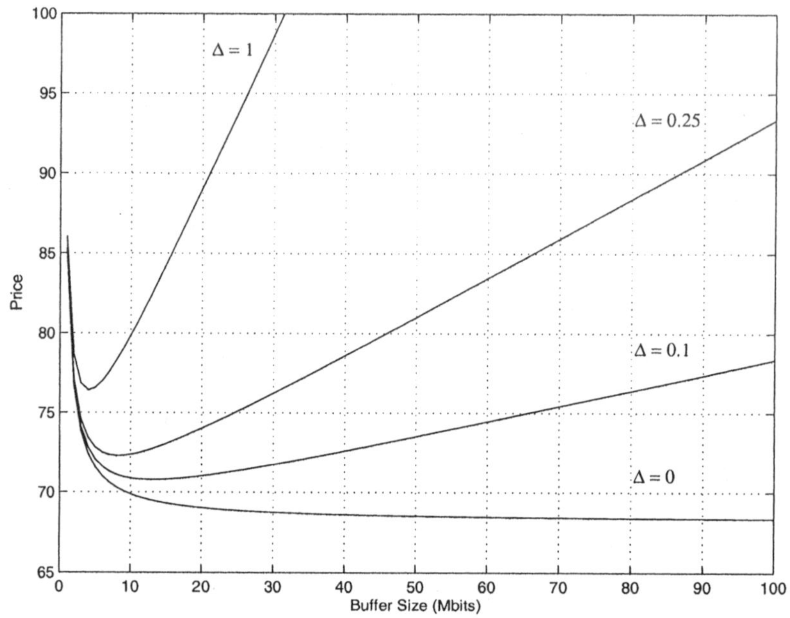

Figure 4. The effect of Δ on the calculated price

than the value calculated for C. If there was no scaling constant then the value of B would overwhelm the ex-post pricing component (as shown in Figure 3 for the same trace as in Figure 2) and lead to charges that would be based mostly on the selection of the buffer size and not the resource usage as it is represented by the value of C. Clearly, this was not our intention when we introduced B into the charging model and using a scaling constant Δ allows us to overcome this problem.

3.2.1 Effect of Delta on the Model. By comparing Figure 2 for which $\Delta = 0$, and Figure 3 for which $\Delta = 1$, it should be clear that Δ can significantly influence the shape of the charging model. In order to better demonstrate the role of Δ in our model we have calculated the prices that would be generated by our charging model for a specific trace but for different values of Δ for a wide ranging choice of buffer sizes.

Figure 4 shows that when Δ is small or zero the buffer size does not affect the price calculated by the model. (Note, this is for a different trace than used in the previous figures. All traces used in the figures are identified in the Appendix.) However as the value of Δ increases buffer becomes all the more important. Specifically for smaller values of Δ it seems that effective bandwidth is dominating the price while for larger values it is the buffer size that has the major effect. Ideally Δ should have a value that balances the effect of the two parameters that control the price. In the following section we discuss extensively what we consider appropriate values for Δ and how they can be calculated.

3.2.2 Setting Delta. Since Δ was introduced in order to balance the effect of the two resources used by the customer we suggest setting it equal to the ratio of differences of these two resources as follows.

$$\Delta = \frac{|C_H - C_L|}{B_H - B_L} \tag{5}$$

where B_H and B_L are the highest and lowest values respectively that the buffer size can assume for a specific type of link; and C_H and C_L are the corresponding values that effective bandwidth assumes (for the trace) when B_H and B_L are used as the buffer sizes. In our computations (see Section 4) we assumed that the possible buffer size values will be in the range $[0.01R_p, 0.9R_p]$, where R_p is the peak rate of the incoming link (i.e., the link capacity in terms of bps). Actually, the exact bounds of this range are not that critical as long as the actual buffer sizes available to the customer fall into that range.

On further examination of Equation (5) it should be evident that to calculate Δ one needs not only the link capacity and the values selected for B_H and B_L, but also the specific trace for which C_H and C_L are calculated. If the ISP uses the trace that it is applying the ex-post charge to, in order to determine C_H and C_L, then the ISP will be able to calculate Δ only *after* the end of a billing period. However, this is contrary to the ex-post pricing concept where the charging model must be specified explicitly in advance so that the customers will be able to estimate their charges based on the traffic they send over the network (and thus be able to manage their traffic to potentially lower their charges). Also it would be fairer from the customer's point of view if the ISP offers to everyone using a link of a given capacity the same Δ. Moreover selecting a constant Δ will help the customers plan ahead and select an appropriate buffer size for their connection. If Δ were to change frequently (for example within every billing period) then it would be very difficult, if not impossible, for a user to determine the right choice of buffer size as well as to manage their transmissions in order to minimize their charges. Consequently, we impose the condition that Δ has to be constant for a given link type.

In Section 2 we mentioned that in order to calculate effective bandwidth one needs to measure the mean of the burst periods and the utilization (since all the other variables are known). Looking at Equation (5) we see that the only unknowns are the two values of effective bandwidth (i.e., C_h and C_L) in the numerator. As a result, since the value of effective bandwidth (for a given buffer size and packet loss probability) depends on only two variables utilization (ρ) and mean burst-period (b), we see that Δ for a specific link capacity actually depends on only utilization (ρ) and the mean burst-period (b). Consequently, we suggest that they be set equal to the utilization and mean burst period that the provider considers an average "well-behaved" customer would have. This provides a uniform value of Δ for a given link type.

As an example we will now calculate the price that a "well-behaved" customer would have to pay with respect to the buffer size chosen, for an OC-3 link (capacity: 155Mbps). We calculate these prices using the Δ value set with Equation (5). Specifically we assume the "well-behaved" customer has a utilization of 35% and a mean burst period of 3.5E-04 sec. The choice of an acceptable utilization was based on conversations with our contacts in industry. For the mean burst period however things were somewhat more complicated. We mentioned earlier that we treat bursts as consecutive packets that were transmitted in the same millisecond. In the time frame of a millisecond an OC-3 link can transmit 155Kbits. However, well-behaved customers that transmit traffic only 35% of the time will probably send on average: 35%*155Kbits per millisecond. This amount of data (i.e., the burst) will be transmitted in: 35%*155Kbits/155Mbps = 0.35*0.001 sec = 3.5E-04 sec. Several experiments we have done on trace data indicate that the measured mean of the burst periods, that were calculated for different traces, are greatly dependent on utilization and could be in fact be approximated in the above way. So by using the utilization and mean burst period values mentioned we were able to determine Δ and also calculate the price that a "good" customer will be charged.

We plot the price of a well-behaved customer as a function of buffer size in Figure 5 (using the same trace as in Figure 4). One can see that neither effective bandwidth nor buffer size dominate the price. (We note that in the figure there are no specific monetary units associated with the price.) Specifically, we can see that for very small buffer sizes the price is high since the value that effective bandwidth assumes will be significantly large. However, as the customers choose larger buffers they are able to reduce the price they pay since effective bandwidth falls sharply. Nevertheless, choosing a very large buffer proves to be inefficient since buffer also affects the price and as a result there is a high charge associated with that choice. This behavior is in accordance with the points that we made earlier on the characteristics of a good charging model. Customers will want to select a buffer that will correspond to the minimum of the graph so that they can minimize their costs. In order to achieve this, customers will have to be aware of the type of traffic they are sending because as we will experimentally show in the next section the behavior of their traffic will shift the minimum of the curve.

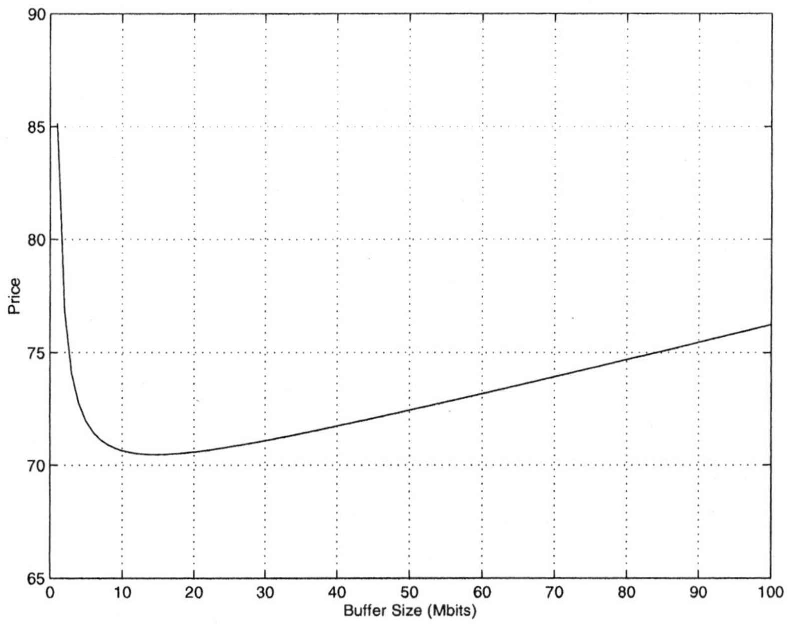

Figure 5. The price calculated for a well behaved customer

4. Numerical results

In order to test the proposed models we used real world traces captured on different networks. The traces consist of the IP headers of the packets captured on specific links. From the IP headers we were able to extract the byte count (i.e., number of bytes of the packet) and the timestamp (i.e., the time the packet arrived at the switch and was captured by the metering software) of the captured packets. These two variables, timestamp and byte count, are the only inputs required for the calculation of the effective bandwidth bound we use.

In the following sub-sections we will present results generated by using traces obtained from the National Laboratory for Applied Network Research (NLANR) repository (http://www.nlanr.net/Traces/). In our experiments we used many of the short 90 second traces provided by the Passive Measurement Project team of NLANR (http://pma.nlanr.net/PMA/) and some of the day-long traces that were captured on the New Zealand to US link. The short traces were originally used to explore the reaction of our models in different contexts while the long traces were used to verify the consistency of our approach in real world settings. Below we present some results obtained with the use of a few selected short traces, in order to demonstrate the behavior of our charging model (the particular trace(s) used in each figure are identified in the Appendix).

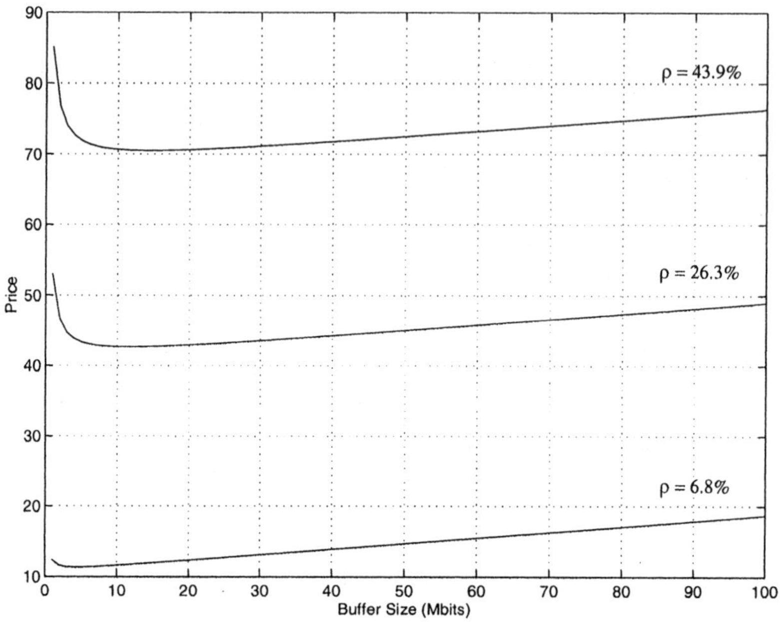

Figure 6. The effect of a customer's utilization on the calculated price

Utilization. Utilization (ρ) represents the percentage of time that the customer is sending traffic over the connection. It is probably the most important of the customer's traffic characteristics. A higher utilization value means that the customers are sending more traffic and consequently they should pay more. Figure 6 presents the results produced by the Delta model for three different traces captured on the same link at different time intervals. The graph shows the price calculated with the same Δ for different values of buffer size for the three different traces. From the graph it is evident that the model performs consistently in the sense that increased utilization is penalized by higher prices.

Packet Loss Probability. Packet loss probability determines the average number of packets that are lost over the customer connection during a billing period. Lost packets can occur because of bit errors that might corrupt the header of the packet. A switch that sees a corrupted header drops the packet because it cannot trust the information in the header. In addition to this, packets might get dropped when they reach a switch and the buffer of the switch is full so there is no room to store the packet. When we discuss packet loss probability in this paper we refer only to the latter case where a packet is dropped because of a buffer overflow. Depending on the size of the link or a customer's specific needs different packet loss probabilities might be requested from the provider.

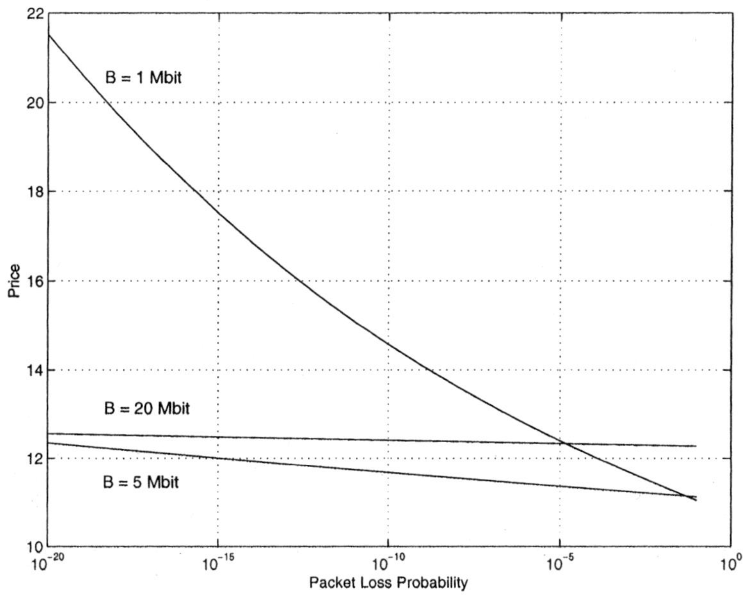

Figure 7. The effect of packet loss probability on the calculated price for a low utilization customer ($\rho = 6.8\%$).

The following graphs provide insight as to how different customers with varying packet loss probability requirements will be charged by our model.

Figure 7 shows the effect of packet loss probability on the calculated price for three different choices of buffer size. We can see that for smaller buffer sizes customers that are requesting lower average packet loss have to pay a higher price. However if the customer has already opted for a higher buffer then the increase is significantly smaller. Figure 8 provides the same information for a different trace with significantly higher utilization, 43.6% as opposed to the utilization of the trace used in Figure 7, 6.8%. For the higher utilization trace the slopes of the lines remain roughly the same. The only difference is the lower price of the 20 Mbit buffer choice with respect to the other two choices. This occurs because for higher utilization the minimum price with respect to buffer size occurs for larger values of buffer (see Figure 6). As a result small values of buffer will generate higher prices.

Price vs. Burstiness. As we have already pointed out customers with bursty traffic should be charged more than customers whose traffic is well behaved. Moreover the model should motivate customers with bursty traffic to select larger buffer sizes since these buffers will make sure that when large bursts occur there will be little or no packet loss.

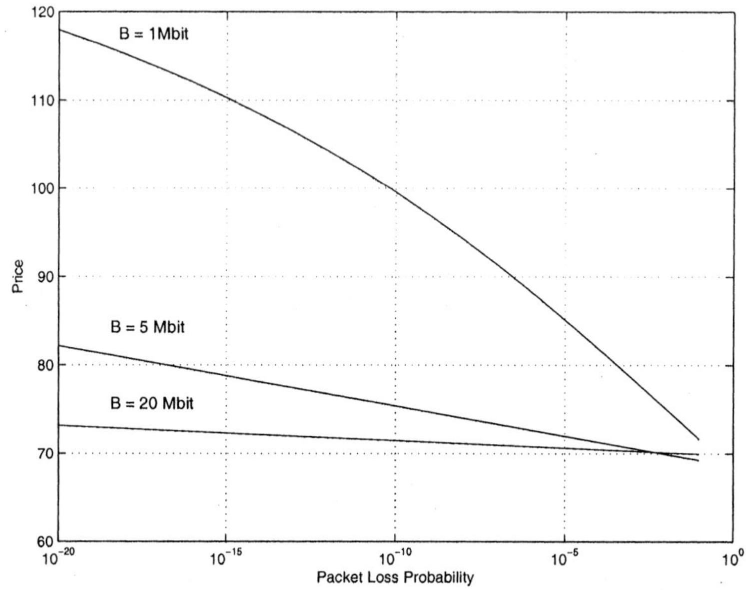

Figure 8. The effect of packet loss probability on the calculated price for a high utilization customer ($\rho = 43.9\%$.)

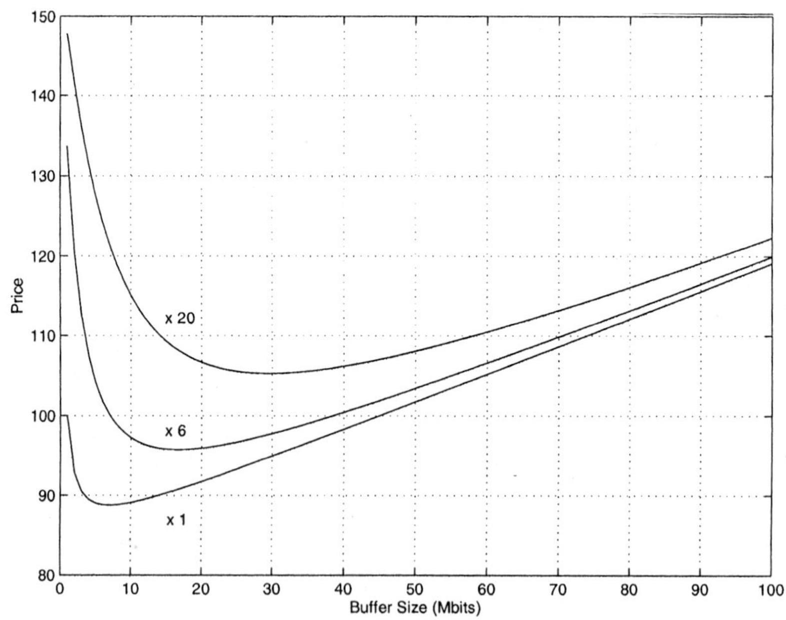

Figure 9. The effect of burstiness on the price curve

Trace	Buffer Size (Mbits)
x 1	7
x 6	17
x 20	30

Table 1. Optimal buffer sizes for different traffic behaviors

Figure 9 shows how the price changes for a specific trace that has undergone special manipulation so that although it's utilization remains constant it's burstiness is increased. In order to achieve this we aggregated consecutive packets so that they appear as if they were one. This way we keep utilization constant since we are not inserting packets to the trace while at the same time we force the packets to arrive at bursts. The more we aggregate the more bursty the traffic becomes. The multipliers in Figure 9 indicate the number of sequential packets aggregated (i.e. "x 1" corresponds to the original trace,"x 6" corresponds to a trace created from the original by aggregating every 6 packets into one and so on).

From Figure 9 one observes that apart from the overall increase in price each curve reaches the minimum point for a different buffer size. As the traffic becomes burstier customers will be directed towards larger buffers in order to reduce their costs. In Table 1 we can see the buffer sizes that correspond to the minimum point for each of the manipulated traces.

5. Final Remarks and Conclusions

By looking at the various experiments in the previous section it is evident that the charging model that we proposed behaves in accordance with the desirable properties of a pricing model that we specified earlier in this paper. It consistently penalizes customers with high utilization and/or bursty traffic and charges higher prices to those who seek better packet loss guarantees for their traffic. Although, for brevity, we presented limited results in the previous section to demonstrate the behavior of our charging algorithm we have actually conducted an extensive set of experiments of a similar nature on a wide range of trace data and for links that varied from 10Mbit Ethernet buses to OC-12 fiber optic carriers. The model has behaved consistently in all of these cases.

To get a better assessment of our charging model we would have liked to test this charging model in practice at an ISP. This will enable us to understand better many of the real-world aspects of implementing this charging model, as well as to observe user (and ISP) behavior in response to such a charging mechanism. We think it likely that an ISP is better able to shape the "long tail" of customer preferences through an ex-post charging approach. Rather than an ISP having to build its network for peak congested periods, the ex-

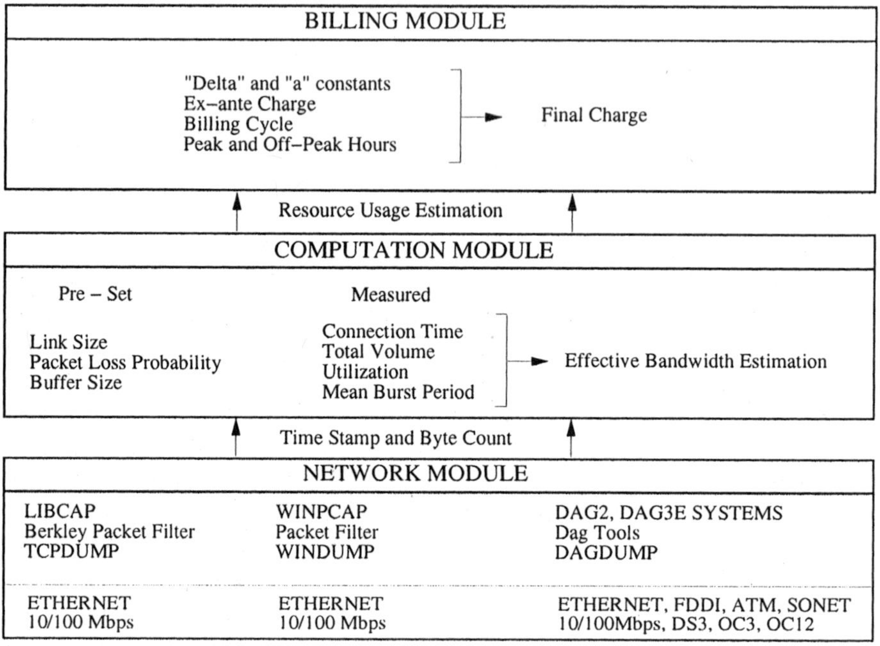

Figure 10. Architecture for implementing ex-post charging at an ISP.

post charge provides an incentive for customers to help shape and manage their network traffic before it is received by the ISP. Although the market has not yet achieved a level of sophistication to immediately implement our ex-post charging mechanism, there is some evidence that service level agreements are more complex and address some of the incomplete contracting problems that had previously existed. To further understand how the ex-post charging model works in practice, we have been discussing our charging model with service providers. As a consequence of those discussions, we elaborate briefly as to how our charging model can be easily implemented by ISPs with no significant cost.

The packet capturing architecture behind our charging model that the ISPs are required to implement can be seen in Figure 10. At the lower level of this diagram the "Network Module" is responsible for capturing the packets and reporting the time they arrived and their size. It's essential components are a network adapter (e.g. a simple Ethernet Card in the case of Ethernet connections or an ATM card in the case ATM connections) that receives the user's packets, an appropriate driver (e.g. tcpdump, windump) that can process all the packets on the physical link and a set of libraries (e.g. winpcap, dag2) that will be used as an interface with the module. Fortunately all these components are readily available and well documented so there is no developing work to be done on the ISP's part. The output that is required from the "Network Module" is the packet size and arrival time of the different packets. However instead of just capturing packets it is possible to filter packets as well. Filtering might be a desirable feature if the ISP wants to charge different prices for different kinds of traffic such as TCP and UDP. In that case the "Network Module" would have to explore a packet's header, determine the protocol used for its transmission and report the appropriate values to the higher modules.

At the middle level the "Computation Module" receives the packet information and is responsible for processing that information to determine a user's connection time, utilization, mean burst period and total bytes send. These are the values that are required for the calculation of the effective bandwidth bound that is required by our charging mechanism. This model can easily be implemented in software and calls upon the network module libraries that we mentioned previously so that it can retrieve all the required information. It is important to note that the values calculated here can be updated continuously with the arrival of every new packet in a Bayesian fashion. As a result there are no extraordinary space requirements. Moreover the actual code that can be used to make these calculations can be only a few lines (depending on the platform and coding language). The output of this module is the estimate (based on the bound) of the effective bandwidth of the captured data trace.

Finally at the higher level of this architecture we find what we call the "Billing Module". This will probably be part of an ISP's usual billing system where necessary information for the billing of all customers is gathered. Mainly this information will be comprised of billing cycles, peak and off-peak periods of the day and/or week and the resource usage scalars (i.e., effective bandwidth and buffer size) for every customer. Once all this information is available the ISP will be in a position to calculate the actual charges of different users.

The elements of the "Network Module" and "Computation Module" can reside on a PC that can be connected to the user's access point (e.g., a switch) with a simple ethernet link. In this configuration the switch that is receiving all of the user's packets can be configured to send a copy of every packet to the SPAN (Switched Port Analyzer) port where the monitoring PC will be connected.[3] In this configuration the monitored link should have a capacity equal or smaller to the capacity of the ethernet link that connects the monitoring station to the switch (if this is not the case then the utilization of the monitored link should be low in order to avoid packet losses). A different configuration is to use an optical splitter to make a copy of everything the user is sending and direct it to

[3]This configuration was used for the monitoring of a 5-day trace at the New Zealand Internet Exchange (http://pma.nlanr.net/Traces/long/nzix2.html).

the monitoring station. Both of these configurations are passive in the sense that they don't intervene with the operation of the network and they don't require additional resources.

We hope to be able to convince an ISP to test out our charging model. If we are indeed successful, then our future research will focus on enhancing our charging models in response to our empirical observations of user and ISP behavior to our charging mechanism. We envision a testbed similar to the Internet Demand Experiment (INDEX) Project (Rupp et al., 1998) which studied the behavior of individual (micro) users of dialup connections in response to usage based pricing. Since INDEX used a university setting for its research, we are hopeful to also use a university network to examine the effect of an ex-post charging model. University networks are a likely candidate because the network administrators may be more open to supporting the research objectives of such a test of the ex-post charging model. A similar testbed for organizational users of high-speed bandwidth connections will go a long way in understanding better many of the practical issues related to usage based pricing, as well as in validating our pricing models.

Acknowledgement:. Support for this research was provided in part by the DoD, Laboratory for Telecommunications Sciences, through a contract with the University of Maryland Institute for Advanced Computer Studies. We are thankful to two anonymous referees for their helpful comments.

References

Bailey, J. P., Nagel, J., and Raghavan, S. (2006). Ex-post internet charging. In McKnight, L. W. and Wroclawski, J., editors, *Internet Services: The Economics of Quality of Service in Networked Markets*. MIT Press. to appear.

Besen, S. M., Spigel, J. S., and Srinagesh, P. (2002). Evaluating the competitive effects of mergers of internet backbone providers. *ACM Transactions on Internet Technology (TOIT)*, 2(3):187–204.

Blumenthal, M. S. and Clark, D. D. (2001). Rethinking the design of the internet: the end-to-end arguments vs. the brave new world. *ACM Transactions on Internet Technology (TOIT)*, 1(1):70–109.

Courcoubetis, C., Kelly, F. P., and Weber, R. (2000). Measurement-based usage charges in communications networks. *Operations Research*, 48(4):535–548.

Courcoubetis, C., Siris, V. A., and Stamoulis, G. D. (1998). Integration of pricing and flow control for available bit rate services in atm networks. In *IEEE Globecom '96*, pages 644–648. London, UK.

Falkner, M., Devetsikiotis, M., and Lambadaris, I. (1999). Cost based traffic shaping: A user's perspective on connection admission control. In *IEEE ICC*.

Guérin, H., Ahmadi, H., and Naghshineh, M. (1991). Equivalent capacity and its application to bandwidth allocation in high speed networks. *IEEE J. Selected Areas Communications*, 9(7):968–981.

IHR (2006). Internet health report http://www.internetpulse.net/. Technical report, Keynote Systems.

ITR (2006). Internet traffic report. Technical report. http://www.internettrafficreport.com/.

Kelly, F. P. (1996). *Notes on effective bandwidths in Stochastic Networks: Theory and Applications Telecommunication Networks*, volume 4, pages 141–168. Oxford University Press. Oxford, UK.

Kelly, F. P. (1997). Charging and accounting for bursty connections, in internet economics. In McKnight, Lee W. and Bailey, Joseph P., editors, *Internet Economics*, pages 253–278. MIT Press.

Low, S. H., Paganini, F., and Doyle, J. C. (2002). Internet congestion control. *IEEE Control Systems Magazine*, 22(1):28–43.

MacKie-Mason, K. and Varian, H. R. (1995). Pricing congestible network resources. *IEEE Journal of Selected Areas in Communications*, 13(7):1141–149.

McKnight, L. W. and Bailey, J. P., editors (1997). *Internet Economics*. MIT Press.

Micheel, J., Graham, I., and Brownlee, N. (2001). The Auckland data set: an access link observed. In *14th ITC Specialists Seminar on Access Networks and Systems*. Catalonia, Spain.

Mills, D. L. (1989). Measured performance of the network time protocol in the internet system. Technical Report RFC-1128. http://www.faqs.org/rfcs/rfc1128.html.

Odlyzko, A. (2001). Internet pricing and the history of communications. *Computer Networks*, 517:493–517.

Odlyzko, A. M. (2003). Internet traffic growth: Sources and implications. Technical report, University of Minnesota. Available at http://www.dtc.umn.edu/publications/publications.php.

Rupp, B., Edell, R., Chand, H., and Varaiya, P. (1998). Index: A platform for determining how people value the quality of their internet access. In *6th IEEE/IFIP International Workshop on Quality of Service*, pages 85–90.

Shenker, S., Clark, D., Estrin, D., and Herzog, S. (1996). Pricing in computer networks: Reshaping the research agenda. *ACM Computational Comm. Review*, pages 19–43.

Siris, V. A., Songhurst, D. J., Stamoulis, G. D., and Stoer, M. (1999). Usage-based charging using effective bandwidths: studies and reality. In *16th International Teletraffic Congress (ITC-16)*.

Srikant, R. (2004). *The Mathematics of Internet Congestion Control (Systems and Control: Foundations and Applications)*. SpringerVerlag.

Appendix

Trace identifier	Figure number
BWY-976126448-1	2,3,9
BWY-20000916	6,7
BWY-20010214	4,5,6,7,8
BWY-20001203	6

Identification of traces from NLANR repository used in this paper.

KNOWLEDGE REPRESENTATION FOR MILITARY MOBILITY DECISION-MAKING BY HUMANS AND INTELLIGENT SOFTWARE:

The Mobility Common Operational Picture Data Model and Ontology

Robin Burk[1], Niki Goerger[2], Buhrman Gates[3], Curtis Blais[4], Joyce Nagle[5] and Simon Goerger[6]

[1]Department of Electrical Engineering & Computer Science, U.S. Military Academy; [2]U.S. Army Engineer Research and Development Center; [3]U.S. Army Engineer Research and Development Center;; [4]Modeling, Virtual Environments and Simulation Institute, U.S. Naval Postgraduate School; [5]U.S. Army Engineer Research Center; [6]Operations Research Center, U.S. Military Academy

Abstract: The U.S. military is constructing a Global Information Grid that provides key software services to complex networks of computers and software clients in an operational theater. Commanders' need for accurate and timely information in support of complex decisions requires that application programs, intelligent agents and humans be able to exchange, analyze, interpret and report information to one another. While interoperability of human soldiers has traditionally been accomplished by the creation of tacit and explicit knowledge through training, construction of software applications and intelligent agents for the GIG requires a standardized vocabulary and semantically rich formalization of common sense knowledge for the various domains of operation spanned by military planning and operations. This formalization is appropriately captured in ontologies which both provide representation vocabularies and facilitate information exchange. Our recent project to define a data model and ontology for the Mobility Common Operating Picture and our ongoing work to support dynamically computed Common Maneuver Networks illustrate the knowledge engineering challenges inherent in a domain where humans have traditionally relied on tacit knowledge to evaluate information as it influences key decisions. Distinguishing concepts that are inherently relational in nature from those that represent object attributes is a key success factor.

Key words: Knowledge representation; ontology; artificial intelligence; formal concept analysis; route finding

1. INTRODUCTION

Knowledge representation in support both of information exchange and of automated reasoning has attracted considerable interest and debate over the last decade or more [Sowa 1998]. As attention moved from the mechanisms of artificial intelligence and computationally complex software to the challenges of capturing common sense knowledge about specific domains [Chandrasekaran et al 1998], a variety of formalisms for knowledge representation have been proposed, ranging from the data formats and shared vocabularies supported by XML to semantically richer formalisms including formal ontologies and a variety of logics [Orbst and Davis 2006].

Ontologies hold significant promise for knowledge representation because they capture task-independent knowledge about the concepts, objects and processes in a domain [Guarino 1995] and have been proposed as the appropriate starting place for architecting information systems of all kinds [Guarino 1998]. As Guarino notes, however [ibid], ontology-building draws on multiple disciplines ranging from cognitive science and linguistics to the disciplines common in the domain of interest, thereby posing methodological complexities even as it facilitates ontology-driven software architectures.

Although semantically much poorer, data models for interoperability of application software are well understood [Salter et al. 1991] and continue to play an important role in distributed and heterogeneous information systems. The importance of information sharing is readily apparent in the domain of planning and executing the movement of military equipment and personnel across some area of ground terrain. Whether in natural or man-made, real-world or digital environments, the ability of human decision-makers, traditional software systems and intelligent software agents (including those embedded within autonomously mobile robotic equipment) to understand the lay of the land, identify and convey the optimal positions of elements within the environment and move or be moved to those positions is essential to accomplishing most military tasks. As a result, mobility decision-making in software systems depends heavily on the ability of system designers to formalize both information in the form of structured data and human knowledge regarding the interpretation and impact of that information for the purpose at hand.

This paper describes some lessons learned regarding both the need for and the methodological issues associated with creating knowledge representations to support information sharing and mobility-related decision-making by human and intelligent software agents. Although it addresses a specifically military issue, we believe it has wider applicability to the design of other intelligent decision tools and robotic equipment.

2. BACKGROUND – THE PROBLEM DOMAIN

Planning for the movement of military equipment in a theater requires collection, analysis and integration of information from a wide variety of sources, utilizing a number of tools ranging from informal decision analysis processes through quantitative network flow and optimization models. Historically, this integration and analysis have been accomplished through human-centric and human-oriented efforts, resulting in documents such as the maneuver plans produced during military operations and their associated maps annotated with symbols depicting planned actions. However, an increasing reliance on joint service operations and a research emphasis on autonomous, intelligent equipment are creating a need for formal representation of both structured information and background knowledge required to characterize mobility and maneuver constraints and to determine desired courses of action.

Mobility analysis, maneuver planning and execution of maneuvers are just a few of the planning, analysis and operational tasks required for military operations. To support integrated joint operations by ground, air and sea forces, the Joint Chiefs of Staff have called for a Common Operational Picture (COP) to be generated and provided to all commands. The COP is presented as a single identical display of relevant information shared by the commands and allows drill-down to supporting data [Joint Chiefs of Staff, 2001]. Generation and dissemination of regularly-updated COP displays is the responsibility of software services provided by the Defense Department's evolving Global Information Grid (GIG).

In addition to the overall Common Operational Picture, the Joint Chiefs have called for generation of a Mobility Common Operational Picture (M-COP) to support operational commanders. The M-COP presents information needed to determine where to maneuver in the battlespace. It must support and in some cases provide the integration and analysis of extensive information about the terrain, friendly and hostile forces, equipment, weather and other factors to facilitate planning and execution of movement by troops, manned equipment and autonomously mobile robotic systems across the battlespace.

Traditionally, mobility decisions have been made by soldiers who plan maneuvers in response to operational orders that specify a task to be performed and the purpose (commander's intent) of that task. As software services on the GIG evolve, however, and as robotic systems are integrated into military operations, both the collection and dissemination of maneuver-related information and decision-making regarding maneuvers will require formal representation of domain knowledge.

In response to this need, the Army tasked a team of technical and military experts from its Engineer Research and Development Center, the U.S. Military Academy, the Training and Doctrine Command and the Naval Postgraduate School with the job of developing a representation of the information required to generate an M-COP in an operational theater. This representation has two dimensions: a logical data model and an ontology for the domain of ground mobility and maneuver.

The data model is intended to facilitate interoperability of existing physics-based models, battle command systems, and discrete event simulations used to study and validate requirements for new Army ground equipment and to explore tactical opportunities provided by new equipment capabilities. The ontology is intended both to ensure a common conceptual understanding of mobility by all personnel operating in a theater and also to facilitate machine understanding of ground mobility and maneuver concepts to achieve greater automation in data processing and more extensive reasoning by software agents in support of human decision-makers.

The M-COP knowledge representation will affect more than just the application services provide by the GIG to commanders. It is also expected to facilitate interoperability of Army Future Combat Systems (FCS) equipment, including autonomously mobile, intelligent vehicles and other equipment. The intent is to provide a single, integrated operational picture of events in a military theater containing timely information tailored for various command echelons and individual users through the creation of virtual links between the information requirements on the user side and information sources on the network side. Those sources may include, in addition to established databases, real-time data from sensors, robots reporting obstacles they have identified along a mobility corridor, soldiers reporting about their observations via handheld devices and reconnaissance data collected by satellites and unmanned aerial vehicles. In addition to gathering, and reporting such data, the M-COP data representation must also support services related to ground vehicle mobility, maneuver planning and maneuver execution, including route planning before operations and automated dynamic route finding in response to evolving conditions on the ground.

3. COMPLEXITY OF KNOWLEDGE ENGINEERING REQUIRED FOR THE M-COP

A notional use case modified from Goerger et al. [2006] illustrates the complexity of information and underlying knowledge which a Mobility Common Operational Picture must capture, express and interpret.

1LT Griffin receives a mission from his higher headquarters (HQ) to conduct a route reconnaissance in order to facilitate the movement of humanitarian relief and supply convoys into the fringes of a troubled region. The route will be used to move medical, food, and fuel supplies as well as engineering equipment for reconstruction of roads and hospitals within the expanded area of operations. Knowing there are several things he must consider in planning his mission and conducting reconnaissance in order to provide HQ with the necessary intelligence it needs to perform successful convoy operations along the intended route, 1LT Griffin starts to develop a reconnaissance plan to ensure he collects all relevant information. He receives a listing of check points and ten areas of interest from the unit's intelligence officer, which he is specifically told to investigate for possible impediments to convoy movement.

1LT Griffin's platoon departs on time from Start Point (SP) Green. It moves in a V formation: two sections overwatch the route from key terrain along the way while the headquarters section traverses the route collecting relevant data. Throughout the mission the teams note any terrain that would be adversely affected by inclement weather such as heavy rainfall.

Shortly after crossing the Line of Departure one of the sections (the C Team) notes 30-foot poles with wire (probably electrical lines) on both sides of the road. Some of the wires droop across the road. Due to limited maintenance in this sector, the wires are not continuous. Some wires appear to be professionally installed while others look as if they are self installed by some of the local inhabitants. The C Team Leader notes the location of these poles and wires since they may come in contact with convoy vehicle antennae.

As the platoon passes Checkpoint 5 on their list, Team A observes a vehicle on the side of the road. The A Team Leader notifies Teams B and C of the vehicle's location so they are prepared to deal with it as they traverse the route. Initially they can only identify the object as a vehicle. Further investigation shows it is sitting completely off the road and unoccupied. Sensing a potential improvised explosive device (IED), 1LT Griffin notifies higher HQ and is told to bypass the vehicle and continue with the mission.

The route crosses a river at Checkpoint 6. The Teams provide overwatch for each other as they cross the bridge. They note the load capacity and general condition of the bridge. This includes any potential sabotage or natural degradation from exposure to vehicle traffic and weather. They also note any potential for the bridge to become washed out during heavy rains or increased flow from upstream conditions. The teams complete their assessment by making a careful check of potential fording sites in the area in the event the bridge is not available for use.

At Checkpoint 7, the platoon encounters a highway overpass. The teams check its height and condition. Additionally, Teams A and B scout alternative routes for oversized vehicles that may not make it under the overpass.

The majority of the road is concrete, but sections are worn away and consist of gravel or packed dirt. Team C notes the location of extremely rough sections and if there are road side areas that could be used for refueling or unscheduled maintenance needs. They also note choke points along the route. These include areas that are possibly too narrow for larger vehicles such as Heavy Equipment Transports or armored vehicles to transit. They check the shoulder of the road for ease of entry/exit of the road network. The team also notes steep drop-offs or extremely rough surfaces that can impede rapid transition to off-road travel.

They also identify key terrain along the route such as high ground surrounding the route where enemy or friendly forces could launch an attack or simply observe movement along the route. Conversely, they assess the fields of fire available to friendly troops that use the route as well as cover and concealment.

Prior to Checkpoint 8, which is a road intersection, the platoon comes upon a second smaller road intersection. 1LT Griffin believes this could be confusing to a convoy commander, especially at night, causing the convoy to turn too soon down the wrong road. He makes careful note to mark this location as a potential navigation challenge. He also sends the updated map information through the appropriate intelligence channels to have the new road added to existing map overlays and future printed maps.

Near the end of the route, the road takes the platoon around a small village. As they did at the river crossing, the teams provide overwatch for each other as they pass the village. They note that there is some kind of festival going on in the center of town. Suddenly they hear the distinct sound of AK-47 fire and immediately train their weapons in the direction of the sound. They watch carefully and realize that the celebration is a wedding party and the gunshots are simply shots in the air, common in this culture. They note the incident for future reference.

Throughout the mission, the scout teams communicate with one another using tactical radio systems. They also maintain communication with their higher headquarters. Prior to departing, the unit signal officer advised them of potential "dead spots" for frequency modulation (FM) communications. They perform radio checks in these locations while conducting the reconnaissance to ensure convoys will be able to maintain communications while on the route. If unable to maintain communications in these masked areas, the platoon must identify locations for retransmission stations to cover these dead spots.

Upon completing the route, the teams return to their home base along the route. This trip is faster since they are already somewhat familiar with the route, but they do notice a culvert that runs under the road just outside the village. They had overlooked that culvert when the gunshots went off. Soldiers from the teams leave their vehicles to take a careful look at the entrances to the culvert as a possible place where enemy forces might hide explosives or launch attacks on convoys. The scouts also identify if the culvert load classification would allow it to handle heavy convoy traffic.

When the platoon returns to its home base, it conducts a thorough debriefing with the squadron S2 (intelligence), S3 (operations), and S4 (logistics) officers. They relay all that they had observed during the mission.

Although this reconnaissance was presented as an actual mission, it could just as easily have been part of a simulation-based training mission. For instance, a unit preparing to deploy may want to have its scouts and engineers rehearse route reconnaissance missions in a realistic training environment. This can be coded into a three-dimensional driving simulator that permits the platoon leader to "drive" the route while transmitting information to its own HQ staff, which then will analyze the data before sending it to a higher HQ. This capability would permit units to train in as realistic a scenario as possible.

The information required for the real-world mission would match that of the simulated mission. To support either the actual mission or the simulation-based training exercise, the M-COP must enable seamless transfer of a wide variety of information between the data sources and decision-makers, including decision support software.

Introduction of autonomous equipment such as unmanned ground vehicles into future convoys will create an even greater need for standardized capture, analysis and sharing of information needed to plan a viable route, traverse the route and respond to unplanned events. This information will include specific data about the physical, social and political environments within which the delivery of supplies and other movements will be made. In addition, if intelligent software agents, either those embodied in robotic equipment or server-based agents providing data services via the GIG, are to make decisions on the basis of such information, they will need to be provided not only data bases but also knowledge bases that allow them to reason about the implications of environmental conditions for mission planning and execution.

4. FORMAL CONCEPT ANALYSIS AND ONTOLOGY DEFINITION

Ontologies capture and formalize knowledge regarding concepts and their interrelationships which are central to a subject area or domain. A variety of ontology definition languages (ODLs) have been proposed, including the Web Ontology Language (OWL) which has gained significant appeal due to its association with the emerging Semantic Web (http//www.w3.org/TR/owl-features). Early ODLs were frame-based, i.e. they supported the definition of concepts as classes of objects with associated attributes, organized in subsumption (superclass / subclass) hierarchies. OWL and other recent ODLs augment object/attribute definitions with description logics, formal ways to specify relationships between objects which are more varied than the class subsumption (subclass/superclass) hierarchies of frame-based ontologies. Description logics capture object and class relationships in terms that are more meaningful and accessible to human experts, generally resulting in much terser knowledge capture than is possible in first-order logic. Because they also map to first-order logic, however, relationships defined in description logics can form the basis for machine reasoning against semantically rich knowledge bases.

Recent theoretical work in ontology development has highlighted the potential value of formal concept analysis (FCA) in the development of rigorous ontologies suitable for machine reasoning [Priss 2006]. FCA applies mathematical lattice theory to extract classes and their associated attributes from empirical data and existing structured databases. The resulting lattices and sublattices define a subsumption hierarchy of concepts and their extents (i.e. the objects and associated attributes) [Ganter and Wille, 1996].

Formal concept analysis focuses on concepts and concept hierarchies organized within a data or domain context. To be usefully formalized, concepts should be non-overlapping at the same level of generality. More specifically, concepts should be able to be organized into a mathematical lattice structure in which arcs represent relationships between concepts and sub-lattices identify sub-concepts in the conceptual hierarchy. In the usual application of FCA, concepts are inferred by the organization of specific objects into classes defined by specific attributes. The resulting lattices of object / attribute contexts map to concepts for the domain in question. Formal concept analysis fits naturally with frame-based ontologies and has been applied fruitfully to the extraction of object/attribute models from large databases and other data collections.

However, our notional scenario above illustrates the limitations both of FCA and of frame-based ontologies for a domain like military mobility. Consider, for example, the vehicle which Team A sees on the side of the road. As data engineers we would probably note that a vehicle is a mode of transportation which has the attributes of wheels or treads, an average fuel mileage, passenger and cargo capacities and so forth.

As knowledge engineers, however, we must capture the soldiers' understanding that in some situations **a vehicle can be classified as an obstacle or even a weapon system** [Melby and Glenn 2002]. In Somalia in the 1990s, junked cars were pushed into intersections and lit on fire to act as barricades. They are currently employed in Iraq as improvised explosive devices (IEDs) remotely detonated or detonated by a suicide bomber. These "new" uses of the vehicle necessitate a method of describing a given vehicle in **terms of its attributes** (two-wheeled, four-wheel, tracked, etc.) as well as in terms of its **specific purpose at some given time** (obstacle, mode of transportation, etc.) so that all parties receiving information about the vehicle conceptualize the same thing.

Thus, the M-COP ontology must capture knowledge about trucks, about obstacles, and about the relationships possible between these concepts, including the circumstances under which a truck is deemed to be an obstacle. In addition, the M-COP must reflect knowledge regarding the implications of a disabled vehicle. For instance, if it is a friendly vehicle that simply needs maintenance assistance, then it may be pushed to the side of the road, scheduled for maintenance, and eventually become a mode of transportation again. However, if it is set on fire it may be a hostile force weapon system, which in turn may trigger a defensive action on the part of the convoy.

Formalizing these relationships requires a vehicle instance to belong to multiple parent classes – awkward at best to define in a subsumption data hierarchy but easily represented by defining the relationships between these classes using description logics in an ontology. The use of description logic in the M-COP ontology means that the ontology can also be easily updated to reflect new uses for vehicles as tactics evolve [Goerger et al., 2006]. In most cases, this can be done with minimal change in data structures by modifying the description logic against which reasoning will occur.

The vehicle-as-obstacle example illustrates the significant advantage of adopting an ontology-based approach to the software that will generate the M-COP. If the relevant knowledge about tactics were embedded as procedural logic, updating it as events unfold would be cumbersome at best and would be likely to miss portions of the extensive tacit knowledge which our notional platoon used during their reconnaissance mission.

5. PROJECT APPROACH AND PRODUCTS

Because of the complexity of the domain and the extent to which tacit knowledge is central to human decision-making for military mobility, the M-COP data representation effort extended over more than two years. Although the overall intent of the M-COP data representation project was clear, the scope of the data to be formalized and the appropriate methods and forms to be adopted were uncertain when the project began.

The team identified four tasks required for creation of the M-COP data representation. Task 1 was to analyze systems, data structures, data formats and Army doctrine in the context of ground vehicle mobility and the Common Operational Picture [Richmond et al., 2005]. An initial slate of categories and features/attributes for the M-COP was produced in a tabular format and a procedure developed for obtaining input and consensus from stakeholders. The team reviewed Army documents, existing data models and emerging concepts and capabilities of the GIG, current and emerging standards. Out of this information, the team developed the following definition which clarified the scope of the project:

Mobility Common Operational Picture (M-COP): A subset of the COP consisting of relevant movement and maneuver data and information shared by more than one command. The M-COP can be tailored for various users and will include data and information for mobility of individual combatants, ground vehicles, and autonomous/robotic vehicles.

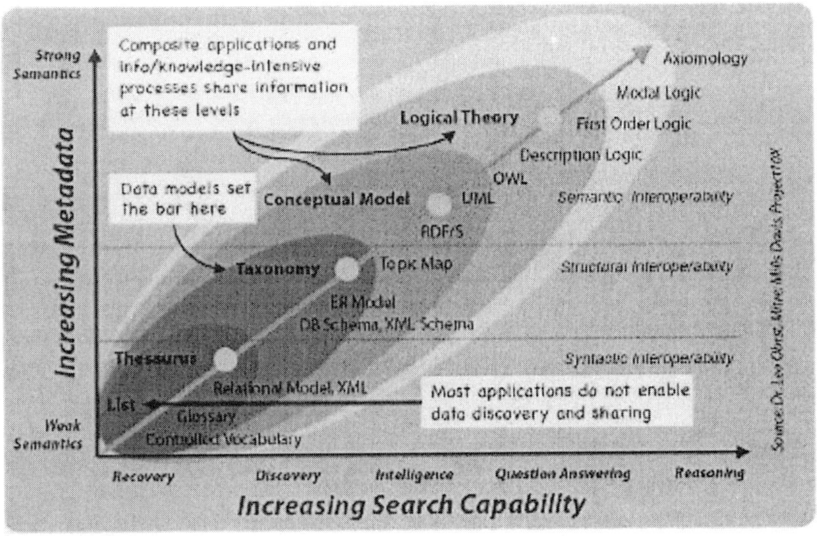

Figure 1. A view of data/knowledge representations (Orbst and Davis, 2005). UML is Unified Modeling Language, RDF/S is Resource Description Framework/Schema, DB is Database, OWL is Web Ontology Language and XML is Extensible Markup Language.

In Task 2 the team conducted a stakeholders' analysis to describe the top-level design of a common data model for M-COP [Blais et al., 2005]. The stakeholders analysis proved to be a valuable method for prompting and collecting expert inputs (knowledge capture) for identification of M-COP information requirements. The inputs provided an excellent foundation for identification of top-level data categories which would guide design of the M-COP data model, along with preliminary identification of software services needed to support M-COP generation in the future GIG environment.

At this point, the team investigated a wide spectrum of data/knowledge representation approaches (Figure 1.) and identified the desirability of achieving as high a representation level for various components of the model as possible within the constraints of the project effort and duration. The team also evaluated a range of data modeling techniques and tools, which provided insights into ontology development to guide model refinement in subsequent phases of the project.

By the end of Task 2 the team had identified principal categories of information requirements of the M-COP, as described in Table 1.

Table 1. M-COP information categories

Categories from Functional Decomposition	Definitions
Terrain	The natural and manmade features and their attributes which may influence mobility or maneuverability of ground vehicles.
Obstacles	Those terrain features or other objects or conditions which disrupt or impede movement of ground vehicles.
Weather	Current and forecasted weather conditions which affect mobility and maneuver (visibility, precipitation).
Maneuver Analysis	The results of an analysis related to ground vehicle movement relative to mission, command and control, local culture and other considerations. Also includes some information classes required for the analysis.
Route Planning	A route plan (directions for moving from A to B), the results of intermediate steps to obtain this plan and some of the required data.
Threat Analysis	The locations, capabilities, and other information (potential actions) relating to things that can threaten mission accomplishment. Note this can include, in addition to enemy forces, local population and cultural effects as they affect friendly maneuver (Melby and Glenn 2002).
Forces	Information relating to maneuver and transportation units, and individual platform locations and capabilities as related to mobility and maneuver.
Utilities	Information (metadata) that may be applicable to all elements of the M-COP.

The most obvious category of information required for mobility and maneuver includes the characteristics of the terrain over which movement will occur. Note that, in addition to spatial characteristics, terrain objects described within the M-COP can have temporal attributes such as an obscurant or contaminated area that disperses over time or a physical feature such as a river bed that can be dry or flooded under certain conditions at different times of the year.

One challenge of the M-COP project was the wide range of concepts involved, from details of the physical conditions on a road surface to high-level concepts regarding the ability of a major unit with heterogeneous equipment to cross a given area of terrain (trafficability). As the range of concepts became clearer, the team defined a high-level entity relationship diagram in Figure 2 to capture the overall relationships required to generate the M-COP.

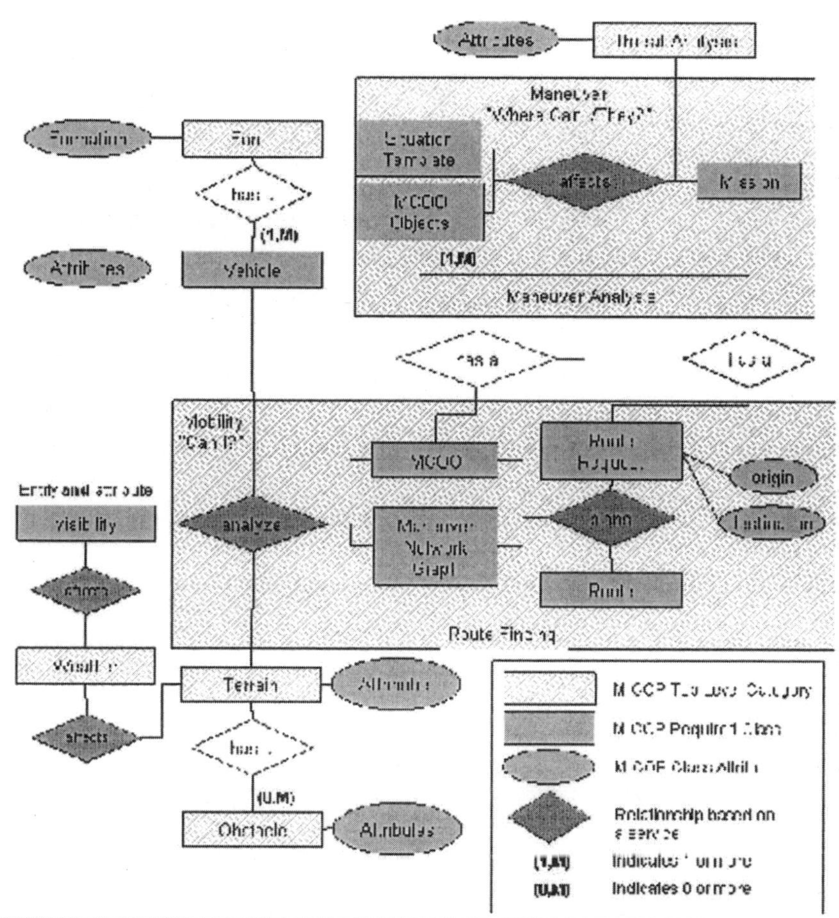

Figure 2. High-level entity relationship diagram for the M-COP.

Task 3 of the M-COP project identified web-based services that the GIG will need to provide in order to generate a mobility common operational picture, along with the data mappings required to provide those services [Richmond et al. 2006]. The M-COP can be viewed as a general service within the GIG that provides data mapping, mediation and storage, and where information based on other data models is interpreted with respect to mobility. It will also depend upon inputs from other services within the GIG.

The intent of the project was to use existing models and formalisms where possible without distorting the overall representation. This was particularly important due to the Army's desire to achieve simulation and command system integration in the near future. As the team investigated existing models (or the lack thereof for some portions of the M-COP data model) and considered higher-order representations, it became clear that a dictionary of M-COP concepts and data elements was required in order to establish a reference vocabulary and to map existing divergent vocabularies to that of the M-COP.

The M-COP data model and formalized semantics have been developed and are evolving in the context of other Army and Department of Defense efforts at standardization. Therefore, during Task 4 the team performed an initial mapping of identified M-COP concepts to the international Joint Consultation Command and Control Information Exchange Data Model (JC3IEDM).

Table 2 shows a very small portion of the data model developed for the Obstacle class. Attributes in this portion are based on the Synthetic Environment Data Representation and Interchange Specification (SEDRIS) Environment Data Coding Standard (ECDS), one of the many existing sources for structured data definitions from which the team drew, where feasible, for the M-COP data model.

Table 2. Attributes for specific Obstacle features (partial list).

Feature Name	Attribute name from EDCS	EDCS Attribute description
BARRIER	COMPLETION_PERCENTAGE	The extent of completion for an object in terms of fractional ascension from start of construction to completion of construction.
	GENERAL_DAMAGE_ FRACTION	The extent of physical injury/damage to an object in terms of fractional degradation from a healthy state. The value may be interpreted as follows: 1/4: Slight Injury/Damage, 2/4: Moderate Injury/Damage, 3/4: Heavy Injury/Damage, 4/4: Fatally Injured or Completely Destroyed.
	HEIGHT_ABOVE_SURFACE_ LEVEL	The distance measured from the highest point at surface level to the lowest point of an object below the surface, as a positive number.
	SURFACE_SLOPE	The maximum slope (rise/run) of the surface of an object.

6. PROJECT CHALLENGES

Although – and to some degree because – of the existing partial data models available, a major challenge for the M-COP team was to choose a starting point for definition of M-COP concepts and data representation. As an example, existing data models are quite detailed for some information such as the natural and manmade features of a location's terrain or, as Table 2 suggests, for features that characterize an obstacle to mobility.

However, in many cases these detailed data models lack an explicit characterization of the relationship between the taxonomy categories and ground vehicle mobility planning and execution, which the M-COP must support. For example, what characteristics make objects in a class such as Road effective for ground vehicle mobility while objects such as Rubble must be avoided? Mobility analysis requires calculation of the maximum speed, acceleration or deceleration possible for a vehicle type over a given area. To support such a calculation, existing terrain features such as dimensions are insufficient. Terrain features must also be described in terms of :

- Off-road: where issues associated with staying on a "path" can be ignored (speed limits associated with path curvature)
- On-road: where path curvature must be considered, and it can be assumed that there is no vegetation to be avoided, or over-run.
- Obstacle crossing: where wet and dry gaps, berms, craters, rubble piles can require additional analysis in which ingress, egress, fording and swimming effects must be considered.
- Obstacle breaching: which can imply operation of engineer equipment such as bull dozers, or mine plows.
- Amphibious operation: assumes ground vehicle swimming; issues are current, wave heights, obstacles, opposing fire, ingress, egress, etc.

Note that these characteristics apply, not to the terrain in its own right, but rather to the interaction of the terrain with equipment which might move across it. As such they capture semantic understanding of the *implications for mobility* of various terrain and equipment characteristics as they interact in the context of a particular task to be planned and executed. This semantic understanding is supplied in current operations by soldiers talking to soldiers, informed by common training. If we are to insert intelligent software agents and autonomously mobile equipment into military operations and planning, however, this knowledge must be formalized so as to allow data interchange and software reasoning against it. In formulating the M-COP data model, the team found that existing Army documents, simulation and data structure standards and data models contain ambiguities and inconsistencies at the conceptual level. We think it likely that this will be the case in any organizational context where tacit knowledge plays a significant role in decision-making.

Exacerbating the team's challenge was the fact that the M-COP's primary purpose is to provide an integrated understanding of the implications of theater conditions on a commander's options for mobility of forces in his area of operation. The higher conceptual levels in the data model, those with the most semantic content, are not "nice to have" add-ons but are central to the creation and purpose of a mobility common operating picture. Yet it was precisely these higher-level concepts which were the least well formalized in existing systems and documents.

The first response of the team to this challenge was to attempt a "top-down" ontology and data model that reflected the diagram in Figure 2, in the expectation that as concepts were decomposed they would eventually map onto the detailed data definitions that currently exist for such domains as terrain.

A stumbling block to the top-down approach lay in the fact that some of the concepts which had been identified by stakeholders as critical to the M-COP's value are not standardized terms found in existing military standards documents and no existing data models were discovered for these concepts. The M-COP top-level category Maneuver Analysis is an example of such a concept. No detailed discussion or standard operating procedure exists that neatly describes "how to" do a maneuver analysis. However, there are several processes and analyses the Army describes that are associated with the Maneuver Analysis category. Officers receive professional training that instills knowledge regarding the nature and scope of maneuver analysis within the prevailing doctrine at the time when training was received.

Thus the team was faced with the need to carefully define and bound concepts which have no official sanction or consistent definition at the moment, but which in tacit or semi-explicit form are key organizing concepts at the knowledge level.

7. FORMAL CONCEPT ANALYSIS APPLIED TO THE M-COP

Given the lack of documented definitions for certain concepts for mobility or of suitable databases from which to extract them, it was not obvious that formal concept analysis would be of much value to the M-COP project team. After a series of attempts to identify agreed-upon higher level classes that would kick off ontology building, however, we revisited that assumption. Could, we wondered, the techniques of FCA suggest a fruitful way to begin?

After reviewing the principles of FCA, we began a series of working meetings in which the team members attempted to define context definitions for their respective sub-domain areas within the M-COP. In some cases, as with terrain, this proved relatively easy since the team was working with existing data models, or at least well defined understandings, and with physical objects and attributes.

However, attempts to define context matrices for sub-domains such as Maneuver Analysis were less successful at first. This sub-domain deals with mostly intangible entities related to judgments and conclusions rather than directly to physical entities like terrain elements. Attempts to build a hierarchy of concepts for this sub-domain proved frustrating and circular. With reference to the principles of formal concept analysis, the team redoubled its efforts however and began working to identify the more concrete concepts and attributes first. In effect we were identifying the lower edge of the concept lattice rather than the upper conceptual edge despite the fact that these sub-domains often centered around non-physical entities. For Maneuver Analysis, for instance, we began with the higher-level concepts identified for the Terrain category, thereby linking the two sub-domains.

This effort immediately identified concepts whose appropriate place in the hierarchy was unclear. For instance, is "trafficability" an attribute of Terrain? The stakeholders certainly spoke of the trafficability of an area of terrain. However, a given terrain is trafficable not in some inherent or independent sense but by a given unit with certain equipment given the weather and surface conditions at the time and whether or not any obstacles were identified on that terrain which the equipment in question could not by-pass.

Formal concept analysis did prove useful for sub-domains like Terrain and in suggesting a strategy of moving from concrete objects to intangible ones for other sub-domains. However, trafficability illustrates nicely why we found it necessary to go beyond the object/attribute emphasis of formal concept analysis for the M-COP. Although "trafficability" seems at first to be an important concept within Maneuver Analysis, in formal terms it is a ***relationship*** between terrain, weather and equipment at a given point in time. Attempting to define it as a class or an attribute merely propagates awkward and unworkable superclasses which have no other reason to be present in the ontology.

With this insight, we were able to make considerable progress. Due to project deadlines for reporting initial data representations, the team decided to concentrate first on data models but to note relationships which would later be expressed using description logic in an OWL-based M-COP ontology. The result at present is an extensive data model plus an emerging ontology in progress as of this writing.

The utility of the M-COP data model and ontology will be tested in the coming year as it is applied to the creation of a Common Maneuver Network (CMN) decision service for route finding across complex terrains for the movement of units consisting of a wide variety of equipment and vehicles.

3. SUMMARY

Providing a Mobility Common Operational Picture to human and machine-based decision-makers requires knowledge formalization, information fusion and high-level information services in a complex, multi-layered environment. As such, it is representative of the kinds of efforts for which formal knowledge management is expected to bring large benefits.

As the Army and the other military services make considerable progress towards deployment of the Global Information Grid and of intelligent robotic equipment, formalized knowledge bases in the form of ontologies incorporating description logic as well as class/attribute definitions will be a key mechanism both for the creation of a common operating picture and also for its use in decision-making. The M-COP data and semantic model is of immediate use for interoperability of models, simulations and battle command systems as the Army continues to develop requirements, concepts of employment and tactics, techniques and procedures involving unmanned ground vehicles. It will also provide an important insight into the requirements for integrating autonomous equipment into the battlespace for operational advantage. In the next year, we plan to exercise this data and semantic model in an ongoing project whose goal is to establish common maneuver networks usable across battle command systems and embedded training and mission rehearsal systems.

As the M-COP project demonstrates, knowledge formalization in support of decision-making which draws on tacit knowledge when performed by humans is a considerably more subtle issue than standardizing formats for structured data. Distinguishing concepts that are inherently relational in nature from those that represent object attributes is a key success factor in that formalization.

9. REFERENCES

Blais, C., Goerger, N., Nagle, J., Gates, B., Richmond, P. and Willis, J. (2005). Stakeholders Analysis and Design of a Common Data Model for the Mobility COP. Project No SIMCI-2005-007. U.S. Army Engineer Research and Development Center, Vicksburg, MS, 31 December 2005. ERDC LR-05-02.

Chandrasekaran, B., Josephson, B., and Benjamins, V. R. (1998). What Are Ontologies and Why Do We Need Them? *IEEE Intelligent Systems and Their Applications*, **14**, 1, 20-26.

Ganter, B. and Wille, R. (1999). *Formal Concept Analysis: Mathematical Foundation*. Springer Verlag, Berling.

Goerger, N., Blais, C., Gates, B., Nagle, J. and Keeter, R. (2006). Toward Establishing the Mobility Common Operational Picture: Needs Analysis and Ontology Development in Support of Interoperability. Paper 06S-SIW-044, *Spring Simulation Interoperability Workshop*, Simulation Interoperability Standards Organization, Huntsville, AL.

Guarino, N. (1995). Formal Ontology, Conceptual Analysis and Knowledge Representation. *Interntional Journal of Human and Computer Studies*, **43**(5/6): 625-640.

Guarino, N. (1998). Formal Ontologies and Information Systems. In N. Guarino (ed.), *Formal Ontology in Information Systems, Proceedings of FOIS'98, Trento, Italy, 6-8 June 1998*. Amsterdam, IOS Press, pp. 3-15

Joint Chiefs of Staff: Doctrine for Joint Operations, JP 3-0, 10 September 2001. Available at: http://www.dtic.mil/doctrine/jel/new_pubs/jp3_0.pdf

Melby, J. and Glenn, R.. (2002). Street Smart: Intelligence Preparation of the Battlefield for Urban Operations. Rand, Santa Monica, CA.

Obrst L. and Davis M. (2006). Ontology Spectrum, from *2006 Semantic Technology Conference Brochure*. Used by permission via personal e-mail communication with Dr. Obrst, 15 December 2005.

Priss, U. (2006) Formal Concept Analysis in Information Science. In: B. Cronin (ed.) *Annual Review of Information Science and Technology, ASIST,* 40.

Richmond, P., Willis, J., Blais, C., Goerger, . and Nagle, J. (2005). *Synthesis of Data Representations for Ground Vehicle Mobility and Suggested Representation of the Mobility COP.* Project No SIMCI-2005-007. U.S. Army Engineer Research and Development Center, Vicksburg, MS, 31 July 2005. ERDC LR-05-01

Richmond, P., Blais, C., Nagle, J., Goerger, N., Gates, B. and Willis, J. (2006). *Web Services Identified for the Mobility-COP.* Project No SIMCI-2005-007. U. S. Army Engineer Research and Development Center, Vicksburg, MS, 1 February 2006. ERDC LR-06-01.

Saltor, F., Castellanos, M., and García-Solaco, M. (1991). Suitability of datamodels as canonical models for federated databases. *SIGMOD Rec.* **20**, 4, 44-48.

Sowa, J. (1998). *Knowledge Representation: Logical, Philosophical, and Computational Foundations.* PWS Publishing Co., Boston.

Printed in the United States